THE
SHIP,
THE
SAINT,
AND THE
SAILOR

THE LONG SEARCH FOR THE
LEGENDARY *KAD'YAK*

BRADLEY G. STEVENS

T0266061

ALASKA
NORTHWEST
BOOKS®

Cover images: Dudarev Mikhail/Shutterstock.com; Morphart Creation/Shutterstock.com

Library of Congress Cataloging-in-Publication data

Names: Stevens, Bradley Gene, author.
Title: The ship, the Saint, and the sailor : the long search for the legendary Kad'yak / by Bradley G. Stevens, PhD.
Description: Berkeley : Graphic Arts Books, [2018] | Includes bibliographical references.
Identifiers: LCCN 2018008242 | ISBN 9781513261379 (pbk.) | ISBN 9781513261386 (hardcover)
Subjects: LCSH: Underwater archaeology--Alaska--Spruce Island (Kodiak Island Borough) | Underwater exploration--Alaska--Spruce Island (Kodiak Island Borough) | Shipwrecks--Alaska--Spruce Island (Kodiak Island Borough) | Kodiak Island Borough (Alaska)--Antiquities. | Alaska--History--To 1867.
Classification: LCC CC77.U5 S75 2018 | DDC 930.1028/04--dc23
LC record available at https://lccn.loc.gov/2018008242

ISBN 9781513261379 (paperback)
ISBN 9781513261386 (hardbound)
ISBN 9781513261393 (ebook)

Proudly distributed by Ingram Publisher Services.

Printed in the U.S.A.

Alaska Northwest Books
is an imprint of

GRAPHIC ARTS BOOKS

GraphicArtsBooks.com

GRAPHIC ARTS BOOKS
Publishing Director: Jennifer Newens
Marketing Manager: Angela Zbornik
Editor: Olivia Ngai
Design & Production: Rachel Lopez Metzger

CONTENTS

DEDICATION

This book is dedicated to the memory of Captain Gary Edwards,
of the F/V *Big Valley*,
who was lost with his ship in the Bering Sea, January 15, 2005.

"May There Only Be Beautiful Things"

ACKNOWLEDGMENTS

THE EVENTS DESCRIBED IN THIS book depended on the efforts of many people without whom the *Kad'yak* would not have been found nor its history revealed. I am greatly indebted to each of them for their assistance and support, and in order not to attach particular emphasis to any one individual, I list them here in more or less chronological order.

Katherine Arndt started the gears in motion by translating the log of Captain Illarion Arkhimandritov's circumnavigation of Spruce Island. Mike Yarborough then planted the seed in my brain by providing me with those documents that both taunted and challenged me for a decade. Bill Donaldson, Mark Blakeslee, and Dan Miller assisted me in some close but misdirected early attempts to find the *Kad'yak* with a two-person submarine and a remotely operated vehicle. Dave McMahan elevated my armchair doodling to real archaeology, and instigated the actual search by providing his support and introducing me to Tim Runyan and the ECU crew. Dave also ran interference between us and the Alaska State Department of Natural Resources, Office of History and Archaeology. Josh Lewis and Steve Lloyd provided key elements needed for the discovery including a boat, a magnetometer, and their time. Bill Donaldson and Verlin Pherson aided the initial efforts by assisting with diving and scuba support. Stefan Quinth documented the discovery efforts on film and later in his book on Kodiak; I believe his presence as a neutral observer helped prevent a difficult situation from becoming worse. Stacey Becklund and the Kodiak Baranov Museum were strong supporters of the search and later exploration efforts. Mary Monroe, as Chairman of the Baranov Museum Board, established a small research fund for our assistance. Marty Owens provided vessel support for the scouting dives in February 2004. Dr. Tim Runyan wrote and managed the grants that helped us to conduct an archaeological investigation once the wreck had been discovered, supervised the ECU team, and taught me valuable lessons about archaeology and the public interest. Frank Cantelas acted as chief archaeologist for the investigation and wrote the subsequent report for NOAA; his calm and contemplative demeanor defused even the most exciting moments. Evgenia Anichenko conducted invaluable research on the origins of the *Kad'yak* for her MA thesis, and she and Jason Rogers formed a significant part of the archaeological team that investigated the wreck. Their booklet published by the Anchorage Historical Museum in Russian and English is the most definitive history of the *Kad'yak* to date and provided source material for this book. Steve Sellers, ECU Dive Safety Officer, insured that our diving met the standards of the American Academy of Underwater Scientists (AAUS), and looked after our personal safety as well. Tane Casserley, on loan from the NOAA Maritime Heritage Program, took valuable underwater and above-water photographs of the 2004 survey expedition, its participants, and artifacts collected for preservation. Captain Gary Edwards recognized the importance of the project and lobbied for the vessel contract; he and his crew made his vessel, the *Big Valley*, our home and dive support ship for the duration of the 2004 survey. Bryce and Jesse Kidd served as crew of the *Big Valley*, whose collective duties included being engineer, first mate, cook, chief bottle washer, plumber, mechanic, general roustabout, longshoremen, crane operator, welder, and other duties as assigned, as well as being ardent listeners, enablers, and supporters of the general expedition welfare. Peter Cummiskey, Larry Musarra, Scott van Sant, Mark Blakeslee, and Sean Weems all assisted our efforts as volunteer divers while surveying the

ACKNOWLEDGMENTS

wreck. Verlin Pherson and Lonnie White provided scuba tanks, air fills, weights, and various other supplies needed by the team. Lydia Black and Gary Stevens provided valuable historical information and documents, without which this story could not have been told. We greatly appreciate the financial support provided by the NOAA Office of Ocean Exploration, Grant #NA04OAR4600043, and the National Science Foundation Grant #OPP-0434280. Captain Craig McLean and Lieutenant Jeremy Weirich, of the Ocean Exploration Program, provided encouragement for our grant requests to their agency. Balika Haakanson and John Adams incorporated our findings into lesson plans on maritime history for students in the Kodiak Island Borough School District. Nicholas Pestrikoff, of Ouzinkie Native Corporation, hosted our visit to Ouzinkie and served as liaison with the Kodiak Native community. Dr. Sven Haakanson, previously Director of the Alutiiq Museum, and now Associate Professor of Anthropology at the University of Washington, answered many questions about the history of the Alutiiq peoples of Kodiak and helped me obtain some historical documents from the University of Washington library. Many of the people listed here read early drafts of the manuscript and helped to correct my errors and omissions, or steer me in the right direction. Olivia Ngai, my editor at Graphic Arts Books, provided many valuable suggestions and insights about organization and story construction. And finally, I have to thank my wife, Meri Holden, and daughter, Cailey Stevens, who participated in some of the adventures described herein, for their patience, encouragement, and ability to quickly burst any bubble of pretense that I might have blown in their direction.

INTRODUCTION

O N A SPRING DAY IN 1861, the *Kad'yak* set sail from Kodiak, Alaska, bound for San Francisco with a shipload of ice. Within a few miles from shore, it struck a rock, foundered, and was abandoned. But it didn't sink. Now a wooden-hulled iceberg, it floated for four days before finally grounding on a reef in Icon Bay, on Spruce Island. That would have been the end of the story but for one detail. Captain Illarion Arkhimandritov, skipper of the *Kad'yak*, had promised to hold a service for Father Herman (now known as Saint Herman of the Russian Orthodox Church) before leaving Kodiak, but the captain did not keep his word. And the *Kad'yak* had somehow drifted through a maze of jagged reefs only to sink directly in front of Father Herman's grave, with the top of the mast sticking out of the water, forming the Russian Orthodox cross—a public rebuke that would forever remind the captain of his perfidy and haunt the site for over a hundred years.

In 2003, after years of painstaking research, I led a team of volunteer divers to discover the wreck of the *Kad'yak*. This is the story about the amazing history of the *Kad'yak* and how it sank carrying Alaska's most important export for two decades in the mid-nineteenth century—ice. Through painstaking research of historical Russian documents and deep analysis of the complicated and confusing log of the skipper who surveyed the wreck site, this is a story of the incredible discovery of a shipwreck over 140 years old, and how I had found it with "friends" who later tried to claim ownership of the shipwreck and credit for its discovery. It is a story about the personal, ethical, and legal struggles to keep the *Kad'yak* safely preserved for the Alaskan people and to illuminate its historical significance and linkage with Alaska Native knowledge.

A true tale of adventure in historical and modern-day Alaska told by a scientist who specializes in underwater research, the story ends with another tragic sinking of the *Big Valley*, the Bering Sea crab boat that served as dive tender and headquarters for the *Kad'yak* expedition, along with its captain and crew, in January 2005.

Come with me, if only briefly, to a faraway place and time. To the real Alaska that is not that different from the imaginary one that lurks in our collective subconscious. Wild, snow-covered in winter, and emerald green in summer, Kodiak Island is central to this story. Approaching it from the fog, Alaska's Emerald Isle suddenly appears as the mists of time part to reveal verdant hillsides reaching up to snow-capped mountains.

CHAPTER I

THE
SHIP
AND THE
SAILOR

A FRESH BREEZE BLEW IN FROM the southwest on the morning of March 30, 1860. Wind from that direction was usually a steady 15 to 20 knots; it was good weather for sailing, especially if one's course was southeast on a broad reach, with the wind off the starboard beam. After a long, arduous winter with constant storms blowing in off the Gulf of Alaska, bringing nearly constant wind and rain to the island, it was a refreshing change. Captain Illarion Arkhimandritov looked at the telltales in his rigging, sails flapping gently, as bellwethers of the coming trip. Standing on deck, he checked the sails, the rigging, and the deck arrangements, and made mental notes of items that needed repair. It was a good ship, this one. It was named the *Kad'yak*, after its home port, Kodiak Island, in the Gulf of Alaska.

While the crew stowed the cargo, Captain Arkhimandritov checked his chronometer, wanting to make sure he got underway in time to catch the incoming tide. In most harbors, ship captains timed their departures to catch the outgoing, or ebb tide, to help carry their ship out of the harbor into the ocean. But not in Kodiak.

Here, the flood tide didn't come directly into the harbor. Because the harbor was situated in a narrow channel between Kodiak and Near Island, the tide moved completely through it.

Situated in the northwestern part of the Gulf of Alaska, Kodiak was at the downstream end of the Alaska Gyre. Currents moving counterclockwise around the Gulf swept by Kodiak, creating a slow but steady current to the southwest. During the outgoing tide, this current was accentuated as water funneled in from Shelikof Strait on the north side of the island, building up to several knots in the narrow channels between the islands of the Kodiak Archipelago. They were famous for their torrential currents, and only the most foolish sailor would pass through those channels on the outgoing tide. But on the flood tide, the backflow of water going against the prevailing current created a gentle river flowing to the northeast, into the island channels and out to the Shelikof Strait. It was that current that Captain Arkhimandritov wanted to take advantage of.

Normally, he would depart from St. Paul's Harbor at the small village of Kodiak. This morning, however, he was departing from Ostrov Lesnoi, the wooded island one mile to the east. There, his ship was tied to the pier, while the crew brought the cargo out from the island in small wooden carts. The cargo was precious. It was the primary product being exported from Alaska, and the predominant source of income for the Russian-American Company (RAC). What the crew was carefully moving was ice, cut from a pond on the island of Lesnoi.

The ice was bound for San Francisco. The fastest growing city on the west coast of North America, San Francisco boasted a booming economy fueled by the gold rush. No longer a sleepy little trading port, it had developed a level of sophistication never before seen on this shore of the Pacific Ocean, and its citizens wanted better lives. More specifically, they wanted ice. Its major purpose was practical. Ice was needed for refrigeration, to prevent food spoilage. But its secondary purpose was purely cultural, and somewhat faddish. It seems the gentry of San Francisco had developed a taste for cold drinks. Ice in their whiskey. Ice in their tea. Ice in their mint juleps. And, of course, ice-cold beer. Where else could you get ice for the greater part of the year but in Alaska? So began the lucrative ice trade with the Russian-American Company.

By mid-morning the crew had finished loading the cargo of ice, all packed in between layers of insulating straw, and made the ship

View of Kodiak waterfront, 1893. A schooner (with sails) is tied up in front of the Erskine House, now the site of the Kodiak Historical Society Museum, along with two 3-masted ships, of similar size to the *Kad'yak*, with Russian Orthodox Church at right. *(W. F. Erskine Collection, Elmer E. Rasmuson Library, Document #UAF-1970-28-418, Archives, University of Alaska Fairbanks.)*

ready for sailing. The fore and main topgallants were unfurled and quickly filled with wind. As the *Kad'yak* began to move forward, the topsails were loosened and bellowed out. The captain's breast swelled with pride at the sight; it was as if the ship had come alive, filled with breath, as it began life anew. Within minutes, they were ghosting steadily down the channel past Ostrov Lesnoi and out into the Pacific Ocean. Their course required them to make several tacks within the first 2 miles in order to pass the extensive shallow reef system that reached out almost a mile from the north end of the island and threatened to grab the ships of unwary sailors. As it glided northeast through the channel that morning, the *Kad'yak* probably had all but its royals set to catch the breeze that blew in from the southwest.

BUILT BY HANS JACOB ALBRECHT Meyer in Lubeck, Germany, in 1851, the three-masted barque was purchased by the Russian-American Company in 1852 and put into service in Alaska as the *Kad'yak*. In those days, ships were not built from plans but from the memories and experience of their builders. So although its exact dimensions were not recorded, it was reported to have a capacity of

Top: The *Charles W. Morgan*, a new England whaling ship (ca. 1849), similar in size and sailplan to the *Kad'yak*. *(Photograph 1972-2-59 from Mystic Seaport Organization.)* Bottom: *The Belem*, a three-masted barque of similar size to the *Kad'yak*. *(Source unknown.)*

about 477 tons. It would have been about 132 feet long, with a beam of 30 feet, and a beam, from deck to keel, of about 20 feet. When loaded, its draft was about 14 feet. The hull and keel were covered with "Muntz metal," a mixture of copper and tin which prevented shipworms from attacking the wood.

Not to be confused with the bark—a blunt-nosed, flat-bottomed hull built to sit on the mud of harbor bottoms while loading cargo— the barque was defined by its rigging, or arrangement of sails. Three-masted barques were the epitome of fast trading ships and the most common vessels on the seas in the years after 1850. The foremast and mainmast were rigged with square sails, including mains, topsails, topgallants, royals, and perhaps a skysail on the mainmast. The topsails may have been split into upper and lower, a modification that became popular in the mid-nineteenth century because they could be furled more easily in heavy weather. Inner and outer jibs and a staysail graced the bowsprit, and multiple staysails hung between the masts. But unlike the traditional square-rigged ship, which would have square sails on all three masts, the barque carried on its third mast, or mizzenmast, a fore-and-aft sail, like that on a modern sailboat. In the latter part of the century, the barque form evolved into the longest, tallest, and fastest sailing ships ever built, with four or five masts, commonly known as clipper ships.

Under the command of Captain Bahr, the *Kad'yak* left Lubeck in July 1851 with a full crew, twenty-eight employees of the Russian-American Company, and a priest, as well as unspecified cargo. After visiting the ports of Kronshtadt, Copenhagen, and Hamburg, the ship headed into the South Atlantic, sailed around Cape Horn, and made a brief stop in Valparaiso, Chile, where several of the crew scattered. The *Kad'yak* finally arrived in New Arkhangelsk (now known as Sitka), capital of Russian America and home of the headquarters of the Russian-American Company in Alaska, on May 7, 1852, after an around-the-world trip of nine months.

The ship underwent a series of changes. Since its purchase by the RAC, the *Kad'yak* was put to work carrying cargo between the Russian-American settlements of Sitka, Unalaska, and Kodiak. It first sailed under the command of Captain V. G. Pavlov, then Captain Herman Debur in 1857, and then Captain Rozmond in 1858.

In 1853, the *Kad'yak* made a trading trip to California and Hawaii with Johann Furuhjelm, the chief of port at Sitka (one could consider this the first Hawaiian vacation cruise). Shortly thereafter, the deckhouse of the *Kad'yak* was deemed unseaworthy and removed, replaced by a glass skylight covered by a metal grate.

The *Kad'yak* first carried ice to San Francisco in 1857, and over the next two years it made six more trips. Although it usually carried ice on the southbound trip, it occasionally took furs, fish, timber, and candles, and usually returned to Alaska with a cargo of beef, flour, and other provisions. Most of the trade stayed between Sitka and San Francisco, but the ship also carried freight between Sitka and Kodiak.

It was not until 1859 that the *Kad'yak* came under the command of Captain Illarion Arkhimandritov.

CAPTAIN ARKHIMANDRITOV WAS WELL KNOWN throughout Alaska. Unlike most ship captains, he was a Creole, the product of a Russian father and a Native mother, and was born on St. George Island, one of the Pribilof Islands, way out in the Bering Sea, probably in 1820. His status in society was somewhat below that of a full-blooded Russian. That he was also a ship captain was an anomaly. When Arkhimandritov was seven, his father paid to have him enrolled in the Mission school in Unalaska. A few years later he started going to sea on sailing ships, and at the age of thirteen he was sent to the School of Merchant Seafaring in St. Petersburg to learn navigation. After graduation, he returned to Russian America, where he was required to earn back the investment the RAC had made in him. Early in his career, at the age of 22, he proved his mettle by saving the company ship *Naslednik Alexander* from what should have been complete disaster.

In September of 1842, the *Naslednik Alexander* was sailing back from California to New Arkhangelsk under Captain Kadnikov, with Arkhimandritov serving as navigator. On September 27, the ship was running before a southeastern wind at a comfortable 11 knots. Captain Kadnikov turned the ship over to First Mate Krasil'nikov and went down to his cabin to change out of his wet clothes. But toward evening, the barometer dropped as the wind and rain increased. Suddenly, a rogue wave rolled the ship onto its port side and caused it to pitch sideways to the waves. The first mate and two helmsmen were instantly washed overboard. The main boom, gaff, ship's wheel, binnacle, and lifeboats were lost, and the ship half filled with water. Below decks, the mass of seawater knocked down the cabin bulkheads and pushed Captain Kadnikov back and forth in his cabin among furniture and debris. Arkhimandritov found himself in a similar situation, but he managed to swim through the wreckage and water and escaped onto deck.

Realizing that the captain was trapped and the first mate lost, Arkhimandritov assumed command and ordered the crew to turn the ship close-hauled into the wind. He then sent rescuers to save the captain, who was still yelling orders, but they could not reach him and soon his voice faded away. Only after righting the ship and pumping out the water was the crew finally able to enter the captain's cabin. They found him and a Kolosh crewman dead under a pile of debris. Arkhimandritov ordered both to be buried at sea. After two more days of storm, the winds abated and the ship was finally put in order, with most of the destroyed cargo and provisions jettisoned. The ship arrived safely at New Arkhangelsk on October 5, and the RAC launched an investigation and recorded details of the disaster. As a result of his efforts to save the ship and its crew, Arkhimandritov was awarded a gold medal by Emperor Nicholas I, which was to be worn on the ribbon of the order of St. Anna.

Thus Arkhimandritov's skills as a navigator and cartographer were widely recognized, and in 1846 he was tasked by Captain Tebenkov, then the manager of the RAC and de-facto governor of Russian America, with mapping the coastline of Cook Inlet, Prince William Sound, and Kodiak. In 1852, many of Arkhimandritov's original charts were incorporated into the first consolidated set of navigation charts for Alaska, known as Mikhail Tebenkov's *Atlas of the Northwest Coasts of America*, which was engraved in New Arkhangelsk.

For a few years Arkhimandritov commanded the steamer *Aleksander II* on its voyages to the *Aleutian*s and Pribilof Islands. While stationed in New Arkhangelsk in 1852, Captain Arkhimandritov had a dispute with the local priest, which escalated into the priest banishing him from the church and prohibiting him from receiving Holy Communion for seven years. In Sitka, this was the equivalent of excommunication. Arkhimandritov could not participate in community activities or continue to work as a navigator. What he did during that period is unknown, but he did not sail again until at least 1859.

All that time ashore may have made his navigation skills a bit rusty. Arkhimandritov may not have been at the peak of his craft when he undertook the final voyage of the *Kad'yak* in 1860.

PRIOR TO HIS LAST TRIP to Kodiak, Captain Arkhimandritov had dined with Chief Administrator of the RAC Stepan Voyevodsky, the acting governor of Russian America, at his home in Sitka. Before leaving, Mrs. Voyevodsky had made a request of the captain. She was a religious woman, devoted to the Russian Orthodox Church, and especially fond of the departed Priest Father Herman. Would he please, she had asked, hold a Te Deum for Father Herman the next time he visited Kodiak? Such a mission would have required the captain to make a separate journey to the grave of Father Herman on Spruce Island, say a prayer for him, and leave a small donation for the church. Despite his previous treatment by the Church, or perhaps because of it, Arkhimandritov did not have the same level of religious fervor as she did. However, his devotion to the Company and the desire to keep in good favor with the manager had convinced him to agree to her request. He would visit the holy man's grave, he had promised, and make the offering. But time and work took its toll. Upon arriving in Kodiak, he found himself too busy and soon forgot about his promise.

At that time, Kodiak was mostly treeless, having been swept clean of all vegetation by glaciers thousands of years previously. In order to obtain wood for building and cooking, the Russians had to visit the wooded island, Lesnoi, or the pine island 10 miles from Kodiak island. The trees there were actually Sitka Spruce, and it became known as Ostrov Elovoi, or Spruce Island. But neither Lesnoi nor Ostrov Elovoi had a good harbor; only the village of Kodiak, with its deep channel next to the rocky shoreline, was deep enough to bring a tall ship in close to shore. Regular trips to collect wood were a necessity then, and the place they went most often was the pine island.

Leaving the channel, the *Kad'yak* passed south of Mys Elovoi, or Spruce Cape, at the northeast point of Kodiak Island. As they did so, Captain Arkhimandritov looked northwest to the island of pines. There on the southeast tip of Spruce Island lay Father Herman's remains, buried under a small chapel in the woods. As he watched the island glide by from a distance, Arkhimandritov suddenly remembered his promise. *Oh well*, he thought, *maybe next time.* He had more important work to do, captaining a ship for the Company, than making frivolous trips to satisfy some doting matron's superstitious whims.

He turned away from the sight and checked the sails and seas once more. His eyes doted on the twist of sails, the slight corkscrew formed as each sail was angled slightly more than the one below it; it was as pleasing a geometry as known to man. Satisfied with the weather and the ship's progress, he ordered the raising of the royals to flesh out the rigging, then turned control over to the chief mate and went below into his cabin. It was the chief mate's job to supervise the actual sailing of the ship. He had not only to carry out the captain's orders but to anticipate them as well. If the captain had to tell him what to do, it was a personal rebuke. It also helped that the crew was a good one. The chief and the second and third mates were all Russians, as was the cook. The rest of the crew were Natives of the Koniag tribe. Only Arkhimandritov was a Creole—but he was a legend in Alaska, and the crew trusted him.

Below decks the captain sat at his desk and recorded in the ship's log the exact time of their departure, the time they passed Mys Elovoi, and their course. Then he examined his charts to determine where they ought to be at the next change of watch. He marked it on the chart so that he could check their progress against it that evening. As he worked, he could hear the splashing of water against the hull as the ship coursed along and the clumping about of men on deck, working and shouting to each other. They were sounds so familiar to him, so comforting, that he could tune them out completely yet still hear even the faintest variation that would signal something unusual. As the ship gently surged over the sea surface, he settled into the rhythm of the sea and felt at peace. If anyplace was home to him, it was here, aboard ship, in the Gulf of Alaska.

The next sound he heard was one that he would never forget. It may have lasted no more than seconds but must have felt like minutes to him. It would live in his memory as the loudest, most excruciating, and most horrible thing he ever heard. First he felt it as a bump, then a scrape, then a loud tearing and crunching before it erupted into an explosive cracking sound. He knew instantly what it was, though his brain tried to deny it for a second. The shock jolted him out of his seat. He burst out of his cabin and ran up to the deck, praying silently that it wasn't what he thought. But it was.

The *Kad'yak* had hit a rock. Sailing at a full clip of about 4 knots with all sails unfurled on a broad starboard reach, the ship had run into an uncharted reef just below the surface, not more than a few miles offshore of Ostrov Dolgoi, the long island. Immediately the ship began to list. The wind still filling the sails dragged the ship over even further. Boards continued to groan and crack as the ship twisted sideways, dragging itself across the rock. Men clung to the ship with panic in their eyes. Cargo and supplies stored on deck strained against their ropes, then broke free and fell into the water. In a moment it was over—the ship slipped off the reef, righted itself, and became silent once more.

The captain shouted orders to furl the sails, hoping to keep the ship upright. The sailors woke from their frozen stances and climbed up into the spars to reef in the sails, doing their jobs professionally in spite of whatever worries they may have had about their predicament. Slowly, the ship leveled out and began to drift, bobbing lethargically in the swells. Arkhimandritov shouted more orders, sending men down below to check the damage. Just as quickly, they came back up. The hold was filled with water, and it was getting deeper. The men uttered the worst words any ship captain would ever want to hear. He could still hear them many years later, as if in a time warp, replayed slowly over and over: "The ship is sinking!"

There was only one thing to do—abandon the ship. Arkhimandritov ordered the men to lower the ship's small boats. Fortunately, no one had been injured during the crash, and all hands were quickly at work. Keeping them busy was also a good way to prevent them from getting out of hand or starting to panic. Turning away, he ran back down to his cabin to grab what he could. If he could salvage anything, it would be the tools of his trade: sextant, compass, spyglass, and chronometer. And, of course, the company books. After throwing them all into his seabag, he dashed back out on deck. The only thing he couldn't take was his seatrunk; it was too heavy, and there wasn't enough time. Anything left in it would have to go down with the ship.

Most of the crew were in the boats by now; a few stood by on deck, waiting for him. After the last man entered the lifeboats, Arkhimandritov climbed in, and they shoved off. The mood was somber. For a while they sat still and watched as the ship drifted and sank ever so slowly. It seemed to take forever. No one talked. They thought about their narrow escape, about things left on board, about their lost wages, their expectations for the trip, now all dashed. They thought about how sad it was to lose the only home many of them had known for some years.

Captain Arkhimandritov looked around and caught their mood. He knew he had to do something. "Is everybody here?" he shouted. "Is anyone missing?" No one was; all the men had escaped unharmed and were present in the boats. "Then grab the oars," he ordered, "it's time to go." Reluctant to leave their ship but happy to be alive, they started rowing. As far as they could tell, the captain was still in control, doing his job. Whatever agonies of doubt and self-criticism may have crossed his mind were not something he would share with them or that they could fathom. A few hours later they dragged the boats ashore at St. Paul Harbor, tired and dispirited, but relieved to be on terra firma.

CHAPTER 2

THE
SAINT

THE NATIVES OF KODIAK BELONG to a branch of the Alutiiq people called the Koniag, and the name *Kodiak* comes from the Alutiiq word *Kikh'tak*. As usual in the days of European expansion, the explorers misinterpreted Native place and tribal names, often substituting the general for the specific. When asked who they were, Native Americans would respond by saying, "We are the People" in their local language, whether it was Iroquois, Cheyenne, Lakota, or Klinkit. But to the invaders, these became the names for each specific tribe. The Russians were no different. Upon arriving in Kodiak, they asked for the name of the place. "Kikh'tak," the locals answered, meaning "island." And so, the Natives' word for "island" became the name of one particular island. Over time, its pronunciation was changed to *Koniag*, then *Kad'yak* or *Kadiak*, and then finally *Kodiak*. In the days of colonization by the Russian-American Company, the island was commonly known as Kad'yak. Thus, when a new ship was purchased for use in the Alaskan trade, it was named the *Kad'yak*, in honor of the island that would become its home port.

Russians first came to Alaska in 1741, with the expedition of Vitus Bering. Bering's men were the first to see the continental landmass of the Alaskan mainland, but only the mountaintops of what is now the Wrangell-St. Elias National Park were revealed through the mist. Bering's men never set foot on the mainland of Alaska, but they did go ashore to Kayak Island and the Shumagin Islands west of Kodiak, where they traded with Alaska Natives. Continuing westward, Bering's ship was finally wrecked on an island off the coast of Kamchatka, now known as Bering Island, where he died in December 1741. After spending the winter there, the survivors constructed a small boat from the wreckage of Bering's ship and returned to Kamchatka, which was only a three-day sail away. Of the seventy-six men who started the voyage with Bering, only forty-five returned to Russia, the remainder having died from sickness or scurvy. The survivors brought with them hundreds of sea otter pelts and told stories of vast numbers of sea mammals they had encountered. These stories encouraged other hunters and explorers to follow the Aleutian Island chain east to the American continent. Over the next fifty years, numerous outposts were established, often in association with Alaska Native villages, in order to hunt sea otters and other marine mammals.

The first permanent Russian settlement in Alaska was established by Grigorii Shelikhov, a fur trader from Kamchatka. He chose Kodiak Island as the location for his settlement and set out to establish his colony in 1783 with three ships, one of which was named the *Tri Sviatitelia*, or *Three Saints*. He landed in a bay on the southwest end of the island, now called Three Saints Bay, in August 1784. From the outset, Shelikhov's plans were to create an extensive Russian empire in Alaska. Although Russian Imperial edict prohibited mistreatment of Alaskan Natives, he fully intended to establish his colony by force. At Three Saints Bay, the Russians attacked the Native village and massacred a large number of Kodiak natives at Refuge Rock on Sitkalidak Island. Thus, Russian-American relations were off to a rocky start.

Over subsequent years, though, Shelikhov's views moderated, and he learned that the only way to maintain his colonies was to improve his relations with Alaska Natives, which he did through gifts and better treatment. He admonished his workers for mistreatment of the Natives, and eventually established a school where Native boys (mostly captured) would be taught the Russian language, mathematics, and navigation. Shelikhov returned to Russia in

1786 and eventually hired a local merchant named Alexander Baranov to take over management of his American colony. In 1799, Shelikhov's company was granted a charter by Russian Emperor Paul and became known as the Russian-American Company (RAC). The charter was to last for twenty years and gave the Company a monopoly to extract resources from the Alaskan territories, with the main goal being to hunt for furs from sea otters and sea lions, which would be traded to China for a lucrative profit. At that time, the region was referred to simply as Russian America; the word *Alaska* was not applied to it until it was sold to the United States in 1867.

Alexander Baranov arrived in Three Saints Bay in 1791 to find a struggling outpost populated by Russian fur hunters, or *promyshleniki*. The location was unsuitable for many reasons, including the lack of a deep water bay. In 1792, the settlement at Three Saints Bay was wiped out by a seismic sea wave (what we now call a tsunami), forcing the Russians to abandon it. Searching for a better location, Baranov resettled at the northeast tip of the island where a deep water channel between small islands formed a protected harbor, which they named Paul's Bay, after Emperor Paul. This became the town of Kodiak.

To call the Russians fur hunters was generous; fur slaughterers is more appropriate. Wherever animals with fur existed, the Russians killed them mercilessly until they were wiped out. When Vitus Bering first explored the Aleutian Islands in 1741, there were millions of sea otters, thousands of seals, and a healthy population of Steller sea cows. By the time that Kodiak was settled fifty years later, the sea cows were extinct, and there were precious few seals and otters left in the Aleutian Islands to hunt. During the year after the discovery of the Pribilof Islands, hunters killed five thousand fur seals, but numbers declined so rapidly that hunting was suspended in 1804. Within fifty years of settling Kodiak, there would be no otters left to hunt in Russian America.

The great value of sea otters was based on the density of their pelts. Lacking an insulating layer of blubber, sea otters depended on their fur for warmth, which is denser than any other animal's on earth. For comparison, the densest human hair is found on your average Nordic blonde, at 190 follicles/cm2. The density of sea otter fur is an astounding 400,000 follicles/cm2, over 2,000 times denser than human hair. It was like nothing the Russians had ever seen, and denser than the closest animal to which they could compare it, which was the beaver. Because of this, the Russians referred to sea otters as *boobry morski*, or "sea beavers," and the furs as *miagkaia rukhliad*, or "soft gold". As late as 1868, sea otter furs were valued at $50 each.

Despite the slaughter and near extinction caused by unfettered hunting of sea otters throughout their range, the idea that they could be eliminated was virtually inconceivable to the Russians. Early attempts by Shelikhov to restrain the slaughter were met with derision from Russian authorities, who decried conservation as an assault on individual rights and private enterprise. Baranov even declared that cessation of hunting could have no positive effect and would actually result in the destruction of the resource. Despite evident scarcity of sea otters, Charles Scammon, in his definitive book *The Marine Mammals of the North-western Coast of North America* went so far as to suggest that maybe they had just moved to "some more isolated haunt" where they could remain unmolested. Human capacity to deny the obvious ramifications of our own devastation has not changed much over time.

In Kodiak, the Russians developed a new way of hunting. Chasing after the sea otters in their ungainly wooden boats was exhausting work, and most of the otters escaped. But the Native skin boats, called kayaks, were faster, quieter, and more seaworthy, and the Natives knew how to sneak up on their prey silently. When otters rest, they wrap a blade of kelp around them as an anchor so that they don't float away. Floating there in a kelp bed, barely above the water surface, they are hard to distinguish from the gas-filled bulbs of the bull kelp. It takes a practiced eye to find them in the middle of a large kelp bed. It didn't take long for the Russians to realize that they could catch more otters if they forced the Natives do it for them. Starting with a two-person Native kayak—or *baidarka*, as they called it, meaning "little boat"—they added a third hole in the middle. A Russian overseer could sit in the middle while two Natives paddled the kayak and did all the hunting.

Hunting parties were usually organized by Baranov and lasted for months. The leaders, or Toyons, of every village were required to identify strong men who would participate in the hunt, and up to one hundred kayaks would assemble at St. Paul Harbor at a predetermined date in the spring. One Native would be designated as the *partovshchik*, or foreman, whose responsibility was to dole out supplies of flour, tobacco, tea, and sugar. After a blessing by the priest, the Alutiiq armada would paddle out of the harbor, not to return until the fall. Arriving at a likely site, the kayakers would spread out in a wide arc. When an otter was seen, it usually made a quick dive underwater. The closest hunter would paddle to the location where it dove and hold his kayak paddle vertically as a sign. The rest of the fleet would then form a wide circle around him. After a little while, the otter would surface to get a breath, and the hunter would launch a dart at it, causing the otter to dive again. The circle of kayaks would tighten around the frightened and exhausted animal, and as soon as it surfaced again, it would be assaulted by a rain of darts, usually killing the animal. The hunter whose dart hit closest to the head would be given credit for the kill and could claim it for payment. It was a deadly, efficient way to hunt. Although the Natives were "paid" for their efforts, it was really more a form of slavery than employment.

Russian relations with the Natives were complicated. The Russians were a wild and free-spirited lot, and with few women to hold their attention, they quickly became troublesome. Some took Native women as wives, often by force. The RAC charter required the Company to treat the Natives as subjects of the Russian Empire, which entitled them to fair treatment; they were to be provided with clothing and food, as long as they provided hunting services in return. However, their services were essentially impressed. Up to 50 percent of the men in each village were required to hunt otters, for which they were paid one-fifth the value that a Russian was paid for the same furs. The Natives continued to be governed by their own Toyons, but these had to be approved by the Company manager, who was usually a senior Naval captain. Despite these stipulations, Russian overseers often treated the Native men who worked for them with contempt, forcing them to hunt until late fall instead of gathering foods that their own families needed to survive.

Despite their inhuman treatment of the Natives, the Russians were a religious group, who followed the rituals of the Russian Orthodox Church. Most of them originated from remote Siberian villages where priests were rare. Realizing that any religion was better than none, the Church at that time allowed citizens to perform their own rituals without the presence of a priest. Many of the Russian workers had come from areas of Northern Russia and Siberia, where their religious beliefs aligned and blended with those of Native groups, and over time had incorporated many Nativist traditions into their own. In Russian America, they continued this practice. Russians recognized and feared the power of Native shamans and occasionally sought their assistance with the weather or health issues. Through these interactions with lay Russians, many Alaska Natives converted to Russian Orthodoxy without any intervention on the part of the Church.

In order to tame the spirits of his workers, nourish their souls, and keep them under control, Shelikhov requested the Russian Orthodox Church to send over a lay priest to administer to the spiritual needs of the settlement and for supplies to build a church. In Moscow, this request was reviewed by none other than Empress Catherine II, who decided instead to send an entire Ecclesiastical mission, consisting of five priests led by Father Iosaf, four postulants, and a lay brother known as Father German (or Herman, in English). Most of them came from monasteries on islands in Lake Ladoga, near St. Petersburg, and were considered well suited for the mission because they were accustomed to cold weather and deprivations. Their trip to Russian America required a three-month overland journey from the monastery at Valaam to Kamchatka, followed by several months at sea. They reached their final destination, on Kodiak Island, in September of 1794.

The priests were not prepared for the crude conditions they encountered there. Kodiak was still a wilderness. There was no church; they lived in huts with bare floors, and had little to eat but dried fish. Several of the priests were dispatched to other settlements. Father Iosaf disapproved of the way in which Baranov's men treated the Natives and sent letters back to Russia complaining about it bitterly. After several years of this, the Church decided to elevate

Father Iosaf to Bishop-Vicar of Russian America, which would give him much greater authority that he could use to control the excesses of Baranov and Shelikhov. In 1799, Father Iosaf returned to Okhotsk where he was consecrated and soon set out for his return to Kodiak aboard the ship *Phoenix*, under the American Captain Shields. He never made it back though, as the ship was lost in a storm and sank probably somewhere in Shelikhov Strait.

Of the original mission to Kodiak, only three priests remained. One of them was Father Herman.

Father Herman felt he was destined for this life. He dedicated himself to care for the Russian men's spiritual needs and welfare. But it was also apparent to him that the Natives deserved his attention as well. Compared to most Native Americans, the Koniag Alutiiq had a high standard of living. They were well fed, thanks to plenty of wild fish, and they were well housed; their barabara-style homes sunk into the ground were warm in winter and cool in summer. But in Father Herman's eyes, they were poor, desperate souls in need of salvation and education; it became his mission to convert them to Christianity first, then to educate them. From this point in history we can look backwards and debate the merits of conversion, but education had definite benefits, the most serious of which was that it would allow the Natives to understand how poorly they were being treated by their Russian overseers and how badly they were being cheated in most transactions. Father Herman may not have realized the full ramifications of his new mission, but Baranov did. It was not in Baranov's interest for the Natives to be educated, as it would only make his job more difficult.

Disagreements between Baranov and the Church came to a head in 1801 when Czar Alexander I took the throne. The new emperor's ascension required that all Russian citizens take an oath to him, and to facilitate this, the priests called all Russian and Alutiiq men to Kodiak. Baranov viewed this request as interference with his control of the colony, and vehemently resisted it, insisting that the Natives go out hunting for otters instead. He even went so far as to exclude the priests from the settlement and threatened to ship them out of the colony.

In 1805, the Russian colonies were visited by Nikolai Petrovich

Rezanof, son-in-law to the now-deceased Shelikhov and heir to the Russian-American Company mantle. He had come on an around-the-world trip to investigate conditions in the colony. Rezanof was either an organizational genius or a complete despot, and during his brief visit he reorganized the structure of the RAC and the methods of accounting and payments, and improved the treatment of Natives. He also established a hospital, a system of courts, and a real school, which promptly took in ninety students, mostly boys. Although the monks had been charged with the task of educating the children in order to turn them into productive workers, their only real interest had been in teaching them the Catechism and how to perform as altar boys. At Rezanof's direction, twenty additional students were sent to Father Herman to learn agriculture. Losing their students may have been a strong rebuke to the monks, but to Father Herman, it must have seemed like a recognition of his work to befriend the Natives and of their respect for him.

Despite Father Herman's efforts to educate the children, Rezanof contradictorily wrote to Moscow that the clergy—and Father Herman, in particular—were not doing enough to subjugate the Natives to the Company's needs. This kind of two-faced behavior was typical of Rezanof, and in an act of particular hypocrisy he accused the priests of mistreating the Natives. Unknown to both Baranov and Father Herman, Rezanof had also been pocketing the generous salaries sent by Empress Catherine II to support the Orthodox mission. After totally disrupting the status quo in Russian America, Rezanof set sail for California, where he inserted himself uninvited upon the Mexican governor, got himself engaged to the governor's teenage daughter, then departed for Russia just as suddenly as he had arrived. Fortunately for everyone except Rezanof, he died on his return trip to Russia, before he could have any further impact on the activities of Baranov or Father Herman.

By 1807, Father Herman was in charge of the Orthodox mission in Kodiak. After several years of increasing tension and deprivations, Father Herman decided that he could no longer continue his work among the community of Russians he had come to serve. His only recourse was to leave Kodiak and relocate to nearby Spruce Island.

Father Herman settled on the southeast end of the island near a small community called Selenie, or Settler's Cove, which later became known as Monk's Lagoon, or New Valaam. There, he built a small chapel and, over time, an orphanage and eventually a school for Native children. It was the first Western-style school in Alaska. For the rest of his life, Father Herman dedicated himself to preaching and teaching the Natives, and for this he was highly revered. Assisting him in his efforts was a Native woman whom he called Mary.

He also found a champion in Lieutenant S. I Ianofskii, who became chief manager of the RAC in 1818. Ianofskii also happened to be married to Anna Baranovna, widow of Alexander Baranov, who had died on his way back to Russia in 1819 after being relieved of his duties as chief manager of the colony. Anna was also a full-blooded princess of the Kenaitze tribe from the area now known as Prince William Sound. During an epidemic of disease among the Aleuts, Father Herman tirelessly attended the sick and devoted himself to their healing. Seeing his efforts, Ianofskii ordered additional funding and supplies be sent to Father Herman and the orphans in his care. Later in his life, Father Herman served as protector to Anna Baranovna, who moved to Spruce Island after the death of her husband and was later buried near Father Herman's chapel in 1836.

Although his work among the Natives was enough to earn him a place in Alaskan history, Father Herman is mostly remembered for a singular event. One day there was a terrible earthquake. The Koniags did not know what caused earthquakes, but they knew they were often followed by giant waves (what we now call a tsunami), and they were afraid. They feared that a giant wave would wash ashore and wipe out their village so close to the water. They pleaded with Father Herman to appeal to his god for divine intervention. Father Herman rose to the challenge. Picking up an icon of the Lord, he walked down to the edge of the water in the small cove. Placing the icon on the sand, he stood and announced to the assembled crowd that the water would rise no higher than this icon. To their great relief, it did not. From that time on, the small cove was known as Icon Bay.

Father Herman died in 1836 and was buried beneath his small chapel. Over a century later, in 1970, he was canonized as Saint

Herman, the first Russian Orthodox Saint from the New World. His canonization was based primarily on the "miracle" worked at Icon Bay, when he saved the Natives from an impending immersion. Every year, on August 11, a pilgrimage to Spruce Island occurs. Natives and Russian Orthodox believers travel to Spruce Island where they hold a celebration for Father Herman. Some come from as far away as California or New York to attend. Even now, 170 years after his death, Father Herman still calls the faithful to Spruce Island.

CHAPTER 3

THE
SINKING

THE NEWS SPREAD QUICKLY IN the small village. The *Kad'yak* had hit a rock and foundered, and had been left adrift. The captain and crew had all returned safely. This was the greatest excitement the town had experienced in years. With little else for entertainment, news and gossip was the centerpiece of village life. Everyone wanted to know, and many wanted to see. Soon, small boats were being launched by Natives, Russians, and Creoles. They all wanted to see the *Kad'yak* before it sank. By evening, a small flotilla of boats had arrived on scene. Most were baidarkies (a Russian invention based on the baidarka), sealskin boats, carrying two or three persons, paddled out onto the ocean. Going out to sea in March was generally risky business for any vessel, but the weather was good, and the baidarkies were extremely seaworthy craft. Their design had evolved over centuries of use by local natives, who were quite adept at navigating them in almost any weather.

To the amazement of the small fleet, the *Kad'yak* had not sunk completely. It was awash up to its gunwales, with just the foc'sle and the masts sticking up above the water as it bobbed up and down in the swells like a sleeping whale. It still had enough surface area above water to catch the wind, and it was drifting along on the surface like a giant piece of flotsam. The next day, it was still afloat.

Captain Arkhimandritov came out to see it himself in a small launch. By now it had drifted to the north, under command only of the mild southern breeze. He considered trying to salvage it, but it was of no use. They had no other boats that could be used to tow it. Even if they managed to tie it up to a fleet of baidarkies, they could not overcome the power of the current and wind, pushing it to the north. And if it sank suddenly, it would take them all down with it. Besides, the cargo could not be salvaged. Better then to let it just drift and see where it went. Who knew, perhaps it would wash ashore on a beach where it could be salvaged.

Many wondered why the *Kad'yak* hadn't sunk yet and began to talk of miracles. Arkhimandritov knew better, of course. Though the spring equinox had arrived and winter temperatures somewhat abated, the ocean in March was at its coldest temperature of the year. Surrounded by tons of ice-cold seawater, the cargo of ice had stayed mostly frozen in the bowels of the ship. Kept afloat by its cargo of ice, well insulated in the hold, the *Kad'yak* had become a wooden-hulled iceberg, floating in the Gulf of Alaska. For three days it drifted as the winds held steady, first from the southwest, then the south, and finally from the southeast, driving the ship closer to shore. On the fourth day the ship finally came to rest, grounded out in the shallow waters offshore of Ostrov Elovoi, the Spruce Island. There, the cargo of ice finally melted, and the ship settled into the bottom, becoming a permanent fixture of the island.

More amazing than its four-day return to shore is the location where the *Kad'yak* came to rest. Selenie Bay, or Settler's Cove, was a name in common use for just about any little cove where Russians settled. On Spruce Island alone, there were at least two, maybe three, coves with similar names. But this particular cove happened to be the place where Father Herman had lived, taught, died, and been buried. The *Kad'yak* had come to rest in Icon Bay, or Monk's Lagoon, right in front of his chapel. And when it finally settled to the bottom, only the mainmast remained standing out of the water, with its topgallant spar horizontal and the main yard slightly tilted, forming the shape of the Russian Orthodox cross.

The symbolism of this was not lost on the Russians or the Natives. Was it divine providence or an accident? To top it off,

Arkhimandritov's failure to honor his promise, to make a devotion to the saint, soon became common knowledge. Whether he had told this story to one of his friends or crewmen, the holy cat was out of the bag. Among the faithful in the community, there was only one explanation for the wreckage: Arkhimandritov had failed Saint Herman, and Saint Herman had claimed his ship.

THE SINKING OF THE *KAD'YAK* was a great loss to the Russian-American Company. In a letter to the RAC offices in St. Petersburg, the manager in Sitka could hardly contain his exasperation that a captain as experienced as Arkhimandritov could run onto a rock in such a well-traveled location. Although Alaska was littered with semisubmerged rocks that were a danger to ships, so many ships had sailed in and out of Kodiak Harbor that authorities wondered if it were a new rock that had just recently appeared:

"It is strange, that the inhabitants of Kodiak until now did not notice that the water breaks above this rock, and some maps even show the channel in this very location. Maybe this rock grew just recently. Although I think that in our colonial seas there are many such new rocks. If the bigger ships sailed more often, they would show to us where the passage is clear, and where it is not; but from such discoveries let God protect me!"

The RAC had only ten ships in their fleet, and the *Kad'yak* was the newest and best of them all. It had been built specifically for the ice trade and was worth over eighteen thousand silver rubles. Although the cargo of ice had been insured, the ship had not. The remaining ships in the fleet were mostly converted Navy ships, many of which had seen decades of use. Despite the value of the ice trade, the RAC had fallen on hard times. After the collapse of the fur trade, Moscow was spending almost as much money to support the colony as it was getting in return. Something had to be done. Within the inner circles of Moscow there was talk about revoking the Company's charter, its license to do business. Even worse was talk about selling the colony to the Americans. None of this reached the colony, however, and if the upper echelons of the Company knew about it, they were keeping it a secret. Recognizing this fact, a letter was sent to Arkhimandritov, chastising him for failing to salvage the ship, tow it to shore, or at least anchor it to the reef where it had grounded so that it could be salvaged later. It seems likely that the writer had not spent any time in Kodiak and had no idea what wind, seas, and storms can do to a ship grounded on a rocky reef, nor how difficult it would have been to raise or move it.

Without a ship, Captain Arkhimandritov now had little to do, but his excellent navigational skills would not go to waste. He was soon reassigned to a new duty as the cartographer of the Alaskan coast. At that time, there were no reliable charts of the area, and Arkhimandritov was a natural choice for the job. At the request of the RAC and by orders of Captain Furuhjelm, now Company manager, Arkhimandritov set out to survey the shoreline of Spruce Island and Afognak Island. On June 30, 1860, he set out in three large baidarkas with four Native paddlers and an assistant. The Aleut baidarka was traditionally a single-person craft, though they could sometimes squeeze in a small child or woman lying down in the front of the boat. Two-person kayaks were rare because they were impossible to right if turned over. To perform what we now call an "Eskimo roll" in a kayak takes coordination and practice. It is not possible to achieve that level of coordination between two people when you are upside down underwater.

Three-person baidarkies, therefore, were an invention of the Russians hunters, who sat in the middle seat, supervising two Natives who did all the paddling. Such was the way Arkhimandritov traveled around Spruce Island that summer, over a period of six weeks. During that time, he took many bearings, recorded many landmarks, and kept a detailed journal. On the third day of his journey, he entered Icon Bay. Standing on a small rocky islet, he pointed his compass toward the mast of the *Kad'yak*, which was still protruding above the surface of the water.

"On this bearing," he wrote, "lies the topmast of the barque *Kad'yak*."

These were the last words recorded about the ship. From there, he continued his journey and eventually gave his journal and a report to the RAC. But whatever became of his work is not known. No new map was produced, or if it were, it has become lost. His notes, however, were the property of the RAC and eventually were transferred to the United States when Alaska was sold in 1867. For over a hundred years, the *Kad'yak* was forgotten.

The sinking of the *Kad'yak* was not the end of Captain Arkhimandritov's troubles. Only a year later, he grounded another ship while leaving the port of New Arkhangelsk. Whether he attributed either sinking to the intervention of Saint Herman is unknown, but possibly to atone for his lack of devotion, in 1869 he donated an icon of Saint Nicholas, the patron saint of sailors, to Saint Herman's chapel on Spruce Island. The icon was later moved to the Russian Orthodox Church in Ouzinkie, and sometime in the 1980s, it disappeared. Almost immediately following the *Kad'yak* shipwreck, a navigational aid was placed above the rock on which it impaled itself. This became one of the first navigational markers in Alaska, and the rock later became known as Kodiak Rock on the National Oceanic and Atmospheric Administration ship charts. Although on the surface this appears to be in reference to Kodiak Island, there are so many such rocks that one wonders why this one in particular should bear that name unless it were also a reference to the *Kad'yak*, which met its demise at this location.

CHAPTER 4

SUBMARINES
AND
CRAB SEX

N O LONGER A SLEEPY LITTLE Russian village, Kodiak is now a bustling fishing port, home to several hundred fishing boats ranging from 20-foot skiffs to 150-foot Bering Sea crabbers. A dozen fish-processing plants dominate the waterfront, along with various businesses to support them, including welders, plumbers, and hardware stores. Not far away lie the various bars and restaurants that provide seafaring men with another kind of support, especially when they are not fishing. Kodiak's economic heyday occurred during the king crab boom of the late 1970s to 1980. During that time, fishermen earned a year's salary in a week or two, and most of them were bedecked with the gold jewelry that became de rigueur bling to advertise their success. The bars were teeming and wild, and money flowed through them faster than beer.

By the time I arrived in 1984, however, the king crab fishery had collapsed, most of the boats had been converted from crabbers to trawlers, and it was a much quieter place. Nonetheless, it was still a successful port, handling over 300 million pounds of seafood annually, including salmon, crab, and halibut, worth over $100 million. At one time it was the highest grossing port in the United States, but that title has since passed to Dutch Harbor (Unalaska), to which the Bering Sea fleet has relocated in their pursuit of Pollock, which now forms the basis for the largest industrial fishery in the world. If you have eaten a fish sandwich at McDonald's, you have eaten Pollock. Most of it, though, gets made into surimi, the rubbery-textured, tasteless, fish-paste product that is the basis of artificial crab and used for sushi in Japan and everywhere else. In less than four years, a fishery for the biggest crabs in the world had been replaced by a fishery for fake ones.

APRIL 1991

In the movies, when the wild-haired mad scientist discovers the secret potion that will bring the dead back to life, cure a deadly epidemic, change granite to gold, or stop the rapidly approaching meteor from smashing into earth, he stands defiantly, hands up in the air, and shouts the most exciting words in science: "Eureka! I have found it!"

But that's not exactly how it works.

Usually, the most exciting words in science are: "Hmmm. That's odd." In fact, most great discoveries are the result of *not* finding what you were looking for, but finding something else instead. Something totally unexpected. And it all began with my study of crabs.

I was lying on my stomach inside a two-person mini-submarine called the *Delta*, 600 feet underwater, on the bottom of the ocean in the Gulf of Alaska, and looking out through a 6-inch porthole. The water surrounding us was barely above freezing, and the pressure was over 300 pounds per square inch; if our little steel tank were to spring a leak, we would be instantly crushed. The muddy seafloor was illuminated by lights, but the water was turbid, and I could only see about 6 feet into the gloom. The pilot, Rich Slater, sat slightly above me on a stool with his head up in a steel bubble atop the mini-sub. From there, he couldn't even see the bottom. My face was only about 6 inches above the seafloor, so I had a slightly better view. As we moved slowly through the gloom, I saw some Tanner crabs, which we were trying to study. Then more. Then many of them came into view, all crowded together. Now we were surrounded by hundreds of crabs.

"What the hell?" I thought out loud. "What are these crabs doing?"

Me with a king crab at the Kodiak Fisheries Research Center.

Moments later Rich stopped the sub as we almost ran into a tall stack of crabs that was higher than the submarine. I looked out portholes on both sides of the sub, and all I could see were crabs, hundreds to thousands of them, mounded up on top of each other in haystack-like piles.

Rich and I both stared in awe—we had never seen anything like it before.

For the previous month, I had been diving in the *Delta* with my colleague Bill Donaldson, a fishery biologist with the Alaska Department of Fish and Game, trying to find mating crabs on the bottom of Chiniak Bay, about 10 miles from the town of Kodiak. But instead of being scattered around in isolated pairs as we expected, all the female crabs were gathered into a dense aggregation in the center of the bay. There, they formed themselves into haystack-like mounds, containing hundreds to thousands of crabs. If it were possible, I could have walked 50 yards on top of crabs without setting foot on the seafloor. Our discovery was mind boggling and totally unexpected. In fact, it was the most amazing thing I had ever seen.

It was a great discovery in the annals of crab science (which, admittedly, are not that thick). And for me, it was the crowning point of my professional career so far. I had arrived in Kodiak in 1984, and spent the next six years surveying crab populations in the Bering Sea. It was my job to estimate the abundance of crabs, and calculate the available quota for the fishing industry. All of this was done by dragging the bottom of the ocean with trawl nets from two chartered fishing boats that went out to the Bering Sea every summer for almost three months. But that was just routine work; it wasn't particularly exciting and didn't fill my need to conduct publishable research.

By 1990, I had been doing my job for six years and was becoming restless. I needed a project I could really sink my teeth into. At that time, I became interested in the mating habits of crab. How do they select mates? And more specifically, how large does a male crab have to be before it can mate? This has important implications for fisheries management because the crabs need to mate before they are captured, otherwise they do not contribute to future populations. Fisheries regulations included a minimum legal size for the crab that was based on certain assumptions about the size at which they became mature, but no one had ever tested these assumptions. What size are mating crabs in the ocean? The only way to find out was to go there and observe them, and the only way to do that was in a submarine.

Even I had thought it was a rather outrageous idea. I didn't know how well a mini-sub would work for the job or how we would capture crabs with it, but I was determined to try. Certain people within my agency thought that it was all a great waste of money and that I should be spending my time doing other things. I have learned, over time, to ignore such scoffs, or at least to listen carefully and file away their criticisms while still pursuing my beliefs. Little has been learned by scientists who acquiesced to majority opinion, and if there was one thing I was known for, it was pushing the boundaries of my job description. In 1990, I had learned about the National Undersea Research Program (NURP), a division within the National Oceanographic and Atmospheric Administration (NOAA) that provided funds for such work. It was one of very few opportunities for federal employees to obtain research grants. (Unlike our university colleagues, federally employed scientists cannot apply for many types of grant opportunities offered by the National Science Foundation, Department of Agriculture, National Institutes of Health, or other NOAA-related agencies.) I had decided to apply for the grant, and in the spring of 1991 we had begun exploring the depths of Chiniak Bay with the two-person mini-sub called the *Delta*.

Our research with the *Delta* turned out to be a great success, generating lots of excitement around the Kodiak docks and within a certain segment of the fisheries research community. Submarines and crab sex even caught the attention of the Alaskan news media and led to a few stories in the *Anchorage Daily News*. I was interviewed by the Anchorage TV station (there being only one) and talked about our work with Senator Frank Murkowski on his weekly radio show. With that success under my belt, I couldn't help but think—what else could we do with a submarine that couldn't be done any other way? Could we perhaps use it to find a long-lost, legendary shipwreck?

MAY 1991

Less than a month after our submarine expedition, I received a phone call from Mike Yarborough, a self-employed archaeologist living in Anchorage. He had heard about our submarine dives and wanted to tell me about a pet project of his. He knew about an old Russian ship that sank 150 years ago near the town of Kodiak and was intrigued by it. He briefly relayed the story to me, of Arkhimandritov's unkept promise, the sinking of the ship, and its journey to rest in Monk's Lagoon. Mike had done some preliminary work on the story and had obtained some information from the captain's records that had been translated from Russian for his use. Would I be interested in looking at the information, perhaps even go look for the ship?

Why not, I thought. If I can find a bunch of crabs in 600 feet of murky water, maybe I can find a ship. After all, how hard could it be? Quite hard, as it turned out. That attitude has typified my career as a scientist, whenever faced with an interesting but intractable question. Many times I've regretted asking that question, but this time I didn't.

A week later, a package arrived in the mail from Mike. Inside were a letter from him recounting the story of the *Kad'yak*, the translation of Arkhimandritov's notes, and diagrams of a ship similar to the *Kad'yak*. I read the material over several times. As I did, the hair on my neck stood up. Could the ship still be there? Would any part of it still be intact? And where, exactly, was it? Could I find it? The prospects of such an endeavor filled me with excitement. *Of course I could find it*, I thought. If it's there. If it's still intact. Sitting on the seafloor surrounded by some of the coldest water in the world, some of it surely must still be recoverable. I was intoxicated with the idea. The *Kad'yak* was in my blood, and I had to find it. But how? And when? That was the hard part.

I shared the story with Bill Donaldson. He, too, was intrigued by the story, and together we began to hatch some plans. Could we go over to Monk's Lagoon on Spruce Island for some exploratory scuba diving? It seemed so simple, like we could just jump into the water and there it would be, waiting for us to find it. Maybe we could even take the *Delta* over there and dive for it. Perhaps NURP would fund a trip to go search for it.

Still flying on the wings of our success with the *Delta*, I walked into the office of my supervisor, Bob Otto, and proposed the project to him. Would he support my taking time to work on it? His response was a cold shock that brought me back to earth. It wasn't my job, he said, to go off treasure hunting. This was not what I was paid to do. I was supposed to be doing crab research, and he was not going to let me to spend taxpayers' money on such a wild goose chase. Forget it, he said, not on their nickel at least. My bubble was popped for the time being.

So I put the materials away in a file cabinet and went back to work, planning for next year's research proposal. What were those crab piles all about? Why did they do that? How many crabs were there? I planned to submit my second proposal to the NURP program in September for more work with the *Delta* the following spring. This project would be focused just on crab aggregations, and we would solve the puzzle. In the meantime, I forgot about the *Kad'yak*.

CHAPTER 5

FIRST
LOOKS

MAY 1992: BILL DONALDSON AND I received funding from NURP for more research on Tanner crab aggregations using the mini-sub *Delta*. Because the peak of crab aggregation had occurred in early May of the previous year, we brought the *Delta* to Kodiak for the second time in May of 1992. For ten days we cruised around the bottom of Chiniak Bay looking for crabs. But to our surprise, aggregation and mating had already occurred. It was apparent that the event had ended several weeks previously because there weren't many crabs present, and those we did find had already mated and produced new egg clutches. We missed the party completely. This was a year in which El Niño, the periodic warming of equatorial waters, was very strong. Ocean temperatures in Kodiak were 5°C, almost two degrees warmer than in the previous year. Perhaps that had something to do with why we missed the aggregation event. Having little else to do and with several days of sub time still in our budget, we decided to take the *Delta* and its mothership over to Spruce Island for a little look-see.

The *Delta* was supported by a 120-foot mudboat called the *Pirateer*. These boats were originally designed for the offshore oil industry in Louisiana and had long, open decks capable of carrying loads of pipe to the offshore oil platforms. As the age of oil discovery

Here I'm standing in the hatch of the *Delta* submarine, ca. 1991.

in the Gulf of Mexico evolved into a period of stable oil extraction, the drilling industry wound down and mudboats became available for many other uses. Some of them migrated to the West Coast, where they were converted for fishing and other uses. As it turned out, the *Pirateer* was just too big to get into Monk's Lagoon, so we anchored slightly offshore, just outside the mouth of the bay.

The *Delta* was a workhorse for the marine biology research community. We called it the Volkswagen of the sea because it was simple, small, and convenient. Rich and Dave Slater, the owners, were a great team to work with, and Chris Ijames, the chief pilot, knew every wire and valve inside and out. The *Delta* has made over 4,000 dives and rarely ever had a problem. In all the time I have used it, we only lost one dive due to mechanical problems, but it was fixed within hours. To launch the *Delta*, we would raise it from the deck with a crane, lower it into the water, and tie it to the side of the ship. I would then climb in through a top hatch and lie prone on the floor of the sub, and the pilot would sit on a small stool, above my legs. The hatch could then be closed for the sub to be lowered the rest of the way into the water, and then the motor would start and the sub would move away from the ship. The pilot would open the valves, and the sub would sink slowly beneath the waves. Inside, we breathed air at surface pressure. The hull of the sub prevented it from compressing.

In the back of the sub, a rack of potassium carbonate crystals removed any carbon dioxide that we exhaled, and the oxygen that we used was replaced by a slow bleed from an oxygen tank.

ON THIS DIVE, I LAY in the bottom while Rich Slater piloted the sub. There was a good 15- to 20-knot breeze that day, with 2- to 3-foot chop on the surface. The sub rocked a bit as we descended, but about 20 feet down I could no longer feel any motion. As we sank, I looked downward, wanting to see the bottom. It was always exciting to watch the bottom "come up" to greet you, because you never know what you'll see—maybe a pile of crabs, maybe some interesting rocks, or perhaps lost crab pots you want to avoid. Maybe we'd see something bigger—a shipwreck. On that day, I thought we'd land on top of the *Kad'yak*.

I could see rocks coming up at us and warned Rich. At 45 feet, we passed the top of some pinnacles. The water was so clear; if I looked up I could see the surface water, and looking down I could see the bottom. We settled to the bottom at 90 feet. There, we motored around for a while, but we kept running into steep ridges of rock sticking up out of the bottom. When we couldn't go around them, we tried going up over a few and back down. The seafloor was mostly gravel with small waves in it, about a foot wide; such waves are indications that surface water movements reach this depth and stir up the bottom. After a while, we couldn't move any further, so we went back to the surface.

Although we didn't find the wreck during that dive, we did learn some valuable information: The bottom was mostly gravel, and if the ship had settled there it probably wouldn't sink in very deep. Furthermore, with a draft of 15 feet or less, it probably drifted into the bay over the tops of all those pinnacles. If it settled down on the inside of them, there was no way it could have been dragged or washed back out to deeper water. If the *Kad'yak* had sunk in Icon Bay, it was still there.

1999–2000
Using the *Delta* was great fun, and there was no experience that could compare to lying on the bottom of the ocean in a little

submarine and looking out through the portholes. But it was expensive—the sub and a support ship together cost about $10,000 per day. Most of my grants were in the range of $100,000 or so, which covered about ten days of sub and ship time. If we didn't find what we were looking for in that amount of time, it was a great disappointment. And in some years, it took us a week or more to find the crab aggregations. I needed to find cheaper ways to see the bottom of the ocean. So over the past five years I designed and built several small video camera sleds that we could tow behind the boat, but they didn't give us a live image of the seafloor.

In 1999, I began using a remotely operated vehicle (ROV). This particular ROV was called a Phantom, and it was powered by several motorized propellers, or "thrusters," that allowed it to fly around underwater. It was owned and operated by Mark Blakeslee, a biologist-engineer who ran his own company, Aqualife Engineering, doing various odd jobs that required his unique mixture of skills in biology, engineering, inventiveness, and Rube Goldberg creativity. Mark and I had met in the mid-1980s soon after we both moved to Kodiak and found in each other a similar mix of adventurousness and offbeat humor. Whenever I had a project that required some devilish bit of techno-wizardry, especially if it involved underwater stuff, I would ask Mark to help me tackle it. His solutions were always fun and interesting, occasionally a bit over-the-edge, and sometimes they even worked.

The Phantom ROV was on the end of a long electrical cable, or tether, that sent power down to it and video signals back up to us. In such deep water, the only way to know the exact position of the ROV was by using a Trackpoint system that provided the range and bearing to the ROV. The Trackpoint operated by sending a high-frequency sonar signal out from a transducer suspended beneath the boat. A pinger on the ROV detected the signal and responded. Mark controlled the Phantom by steering it with a joystick and watching a small video monitor to see whatever was in front of it. It was the ultimate video game.

Using the ROV, we saw extremely dense crab aggregations that seemed to coincide with the lunar tide cycle, and over the winter I published my observations in the *Canadian Journal of Fisheries and*

Aquatic Sciences. But I needed more than just one year's observations to confirm anything. So in 2000, five years since my last grant, I submitted another proposal to NURP to ask for two more years of funding for crab research with ROVs and camera sleds.

APRIL–MAY 2001

Because the ROV took up much less space than a submarine, we didn't need a big ship to support it. So this year I chartered the 50-foot fishing vessel *Anna D*, skippered by fisherman Dan Miller, as our research platform. But results were disappointing. We didn't find many crabs, the tides were weak, and by early May, whatever aggregation may have occurred was all over for the year. However, I still had grant money to cover use of the *Anna D* and the ROV for another week. This was an unexpected opportunity, and we decided we would use it to explore some new ground. For three days, Dan, Mark, and I explored the seafloor in Monashka Bay, on the north side of Kodiak. There, we saw colorful fish hiding among the rock-strewn bottom, octopuses skulking off into the gloom, and large halibut exploding out of the silty sediment, but few crabs.

With two days of boat time left, I made a decision. We would go to Icon Bay and look for the *Kad'yak* with the ROV. It wasn't crab science, it wasn't part of my job, and it wasn't specifically covered by my grant. But those were minor considerations. One of the truths about scientific research is that discoveries rarely came about by pure determination or chance. In most cases, scientists made their preliminary observations while working on other projects. Have a little time or money left over from the last project? Why not make some new observations? Maybe they won't amount to much, but it might help steer the course for next year's work. In fact, it is very difficult to obtain funding for grant proposals without some preliminary data to show that your hypothesis is reasonable and proposed methods practical. A cardinal rule is to never to give back unspent grant money; that suggests you didn't budget carefully, and you might not get what you need next time. And it is always necessary to push the boundaries. To try something that challenges expectations. Something that will require learning new skills and information. Something slightly crazy. Why not now?

Mark Blakeslee with remotely operated vehicle *Rosebud*.

For two days, the *Anna D* sat in Icon Bay as Mark drove the ROV around on the bottom. We intentionally anchored the boat as far into the bay as we felt was safe, given the poor chart of the bay and our knowledge that it was full of sharp, rocky reefs. We watched on the video monitor as the ROV passed over sandy bottoms, rocky outcrops, and kelp-covered reefs. Unfortunately, we could not determine exactly where the ROV was at any time. The Trackpoint system worked well in moderately deep water, but in this shallow bay, with many reefs and channels, the signal bounced around a lot, and it was difficult to determine exactly where the ROV was at any time.

As Mark drove, I constantly scribbled notes about depth, distance, and bearing, and made a crude chart of where I thought the ROV actually was. Another problem with using ROVs was that while looking at a video screen, you have no depth perception or scale. Is that rock large or small? Is it close or far away? The narrow view provided by the camera results in tunnel vision. You might pass right by something without seeing it, and if you turn the ROV around, you lose all sense of direction. How far did we turn? Which way are we facing now? In a submarine, you still have a sense of spatial orientation. But staring at a video screen can be very disorienting.

In the end, we could only make a general picture of the bay bottom. There were at least two channels that ran from southeast to northwest into the bay; each had a bottom depth ranging from 50 to 90 feet, and were mostly covered with sand. Running roughly parallel with the channels were rocky reefs that came up to within 20 to 30 feet of the surface. We might also have detected some hidden basins inshore of the reefs that dropped down to deeper depths, but it was difficult to say exactly where they were. We could even have passed over parts of the *Kad'yak* wreckage and not been able to identify it from the ROV video.

By the end of the second day I concluded that using an ROV was not the best way to explore the bay. I would just have to dive there myself and use my own eyes if I was going to see what lay on the bottom, if anything. Dan started up the engine and pulled up the *Anna D's* anchor. Then he put her in gear and began to turn around. As the boat turned to starboard, we heard a loud thunk and felt the boat lurch.

"What the heck was that?" I asked, not wanting to say what I really thought.

"Crap! We hit the reef," Dan said. He threw the engine into reverse then neutral, and ran out of the wheelhouse. Looking over the side, we could see rocks a few feet below the water that we had not seen before. (Maybe these were some of the mysterious rocks that spontaneously "grew" in Kodiak waters, as suggested by our Russian predecessors.) We must have been anchored right next to them. Dan came back inside, opened a hatch, and climbed down into the engine room. Mark and I listened to the clunks, bangs, and curses emanating from below as Dan rummaged around, checking the bilge, various compartments, and valves.

When he reappeared, he seemed satisfied. "No major damage, as far as I can tell. We're not flooding. It must have just been a minor bump." With that, he carefully backed away from the reef, then turned and headed back to Kodiak.

An hour and a half later, we nosed into the dock. As I was packing up my gear, Dan noticed that his engine temperature seemed a bit high. He climbed down into the engine room to check it out. The *Anna D*'s diesel engine was cooled by a keel cooler; coolant circulated from the engine into the keel, where it was cooled by exterior seawater, then returned to the engine. When Dan took the cap off the coolant reservoir, a geyser of seawater rushed out. He struggled to get the cap back on before it flooded the engine room.

"We must have damaged the keel cooler," he said. "That's gonna be expensive to fix. I'm going to have to put the boat in drydock and get it checked out." He shook his head. "I've got a week before my next fishing trip; I just hope I can get it repaired before then."

A few days later, he had the *Anna D* lifted out of the water and discovered that when she hit the reef, welds in the keel cooler had been forced open, letting seawater in. Fortunately, he had made enough money from my charter fees to cover the repairs and still make a profit.

But the whole experience was a bad omen. We had hit a reef in almost the same place where we thought that the *Kad'yak* had sunk. It wasn't the same reef it had hit, but who knew whether this reef had stopped the *Kad'yak* from washing ashore. Maybe the ship had been right underneath us all the time. And certainly we had not sunk the *Anna D*, but if we had hit a little harder, perhaps we might have.

NEVERTHELESS, MY INTEREST IN THE *Kad'yak* did not suffer from this setback. If anything, it intensified. If it was that easy to get to Icon Bay, and the water was that clear and that shallow, it should be simple to go diving over there. Just get a boat and a few divers, and go do it. But I had to know where I was going, otherwise we would just be wasting a lot of time and effort. I pulled my dog-eared folders out of my filing cabinet and once again pored over the translation of Arkhimandritov's log that Mike Yarborough had sent me. I drew lines on navigation charts, trying to trace each waypoint and landmark that had been mentioned. I carefully reconstructed every hour of the captain's journey in June of 1860, trying to see if it would lead me to the *Kad'yak*. But none of it made much sense.

Nonetheless, I began to make plans for an expedition. If I were to go look for the *Kad'yak*, I would need at least a week of time and would have to charter a boat. I couldn't use one of the boats from our lab, since I wouldn't be doing the work on government time. I'd need some new scuba gear, and we'd need some food. At a minimum it would cost about $10,000, probably $15,000. Where was I going to get that kind of money? I had approached the Undersea Research Program with my idea, but it was just too risky for them; after all, there was no good evidence that the *Kad'yak* still existed or was where I thought it was, and I didn't even know where that was. But you don't get anything if you don't ask, so I wrote up my plan as a proposal and began shopping it around. I sent it to *National Geographic*, but they wouldn't fund it unless I could guarantee them that we would find the ship and make an hour TV show out of it. I sent it to Rolex Corporation (yes, they actually fund exploration), but they weren't interested either, despite being a well-known supporter of exploration and adventure. I sent it to several other small nonprofit foundations, all with the same result. None of this is unusual; less than 10 percent of research proposals submitted to the National Science Foundation is successfully funded, and most scientists have to revise and resubmit proposals several times before they get funding. The major hurdle seemed to be that I was not an archaeologist and did not have any credentials for marine archaeological research.

But I also felt that I was missing something. I wasn't convinced of the *Kad'yak*'s location, and it would be impossible for me to convince anyone else unless I was 100 percent certain of my own argument. I still didn't understand how Arkhimandritov had recorded bearings to some landmarks, which, according to my reading of the chart, would have been extremely difficult to see from a kayak. If I could see it from his point of view, perhaps I would understand.

That was it, I thought. I had to go over to Spruce Island in a kayak and retrace his journey.

CHAPTER 6

A VISIT TO
MONK'S LAGOON

AUGUST 2002: ON A LATE summer morning, I set off in a kayak from what we now call Miller Point, at Fort Abercrombie State Park, at the northeast end of Kodiak Island. My wife, Meri, and my twelve-year-old daughter, Cailey, accompanied me in a double kayak. During WWII, Fort Abercrombie held the location of two 8-inch guns, put there to protect the Navy base at Womens Bay, 10 miles to the south. The guns were originally battleship guns and were installed on top of special rotating carriages, on top of a 100-foot-high bluff looking out over Monashka Bay. The guns were never fired, and after the war they were destroyed in place. Now, a military museum occupied what was once the ammunition battery. But if you stood where the gun emplacements were, you could look out across 4 miles of open ocean to Spruce Island and see the entrance to Icon Bay along with some islands. On one of those islands, Arkhimandritov had stood to take his bearing on the *Kad'yak* over 140 years ago. That was our destination.

Paddling a kayak in the open ocean is always a dangerous activity, whether you are 2 miles or 200 feet from shore. Anything can happen, and you need to take precautions. I had never gone kayaking before coming to Kodiak, but Meri was an experienced kayaker and had introduced me to the sport. Every summer we used to make day trips or overnight camping trips in our kayaks. After Cailey came into our lives, those trips became shorter and less frequent, but by the age of five or six, Cailey had her own life jacket and a toy paddle. Now, at the age of twelve, she was capable of handling a standard paddle and doing half of the work in a double kayak.

The kayaks were stuffed to the gills with camping gear: a tent, three sleeping bags, foam mattresses, a camp stove and cooking gear, food for four or five days (because you never know how long you will need to stay), and a first aid kit, plus fishing rods and cameras. Sitting in the kayaks, we were completely covered by spray skirts that hung over our shoulders and snugged up around the edge of the cockpit. We all wore life vests, each one of which carried a whistle and strobe light. In addition, I carried a VHF radio and an emergency position indicating locator beacon, or EPIRB, strapped to the outside of my kayak.

Paddling at about 2 miles per hour, our journey should take us two to three hours to cross the open channel between Kodiak and Spruce Island, depending on winds and currents. The day was partly overcast and there was a light wind from the southeast. Although there was little surface chop, a gentle 2- to 3-foot swell rolled underneath us. We paddled with a steady rhythm, stroking on one side then the other, rotating the paddle in our wrists to get the correct angle and keeping the blades close to the water so they would not drip onto us. With little to focus on except the waves, you could drift into a zenlike state and forget all about regular life, work, and all its associated stresses.

Ten years earlier, I had made this trip with a group of friends. Humpback whales were feeding in the channel then, so we had paddled over for a closer look. For a few minutes the whales were all underwater, so we stopped paddling to see where they would surface. Suddenly, one came up about 100 feet from me, heading straight in my direction, like a freight train bearing down on my kayak. I paddled as fast as I could to get out of the way. But the whale also changed course too and came up again, this time closer to me. I tried paddling away, but it was too late. About 30 feet from my kayak the whale dove, turning his body vertically in the water and raising his flukes up in the air before sinking beneath the waves. The flukes

Getting "fluked" as a humpback whale dives beneath my kayak.

were as wide as my kayak was long, and I felt the water splashing off them. Before I knew it, the whale sounded and glided beneath me. I was stunned. It was an incredible, once-in-a-lifetime encounter, and few would have believed it if it hadn't been photographed by another kayaker downwind from me, who later gave me a copy of the photo.

I thought about that experience now as I paddled and wondered if we would see more whales. I didn't really want to get that close to them again, especially with my family in tow. For most of the trip we kept the kayaks within a few yards of each other, but sometimes we would separate a little farther if one of us took a break from paddling. Occasionally, a swell came up between us, and all I could see was Meri's head above the water as her kayak settled into the trough on the other side of the wave. But it didn't worry me because the water and wind were fair. The middle of the channel was over 600 feet deep. I know, because I had spent a week cruising the bottom of it in the *Delta* back in 1991. As we got closer to Spruce Island and the water shallowed, the swells died down and a light chop arose. About a mile from Spruce Island, Meri woke me out of my paddling rhythm.

"What's a rock doing out here in the middle of the channel?" she asked. "And why is it moving?" I looked to my left where she was pointing and saw a gray pyramid rising from the water. It was indeed moving at the same speed we were.

"That's no rock," I said. "That's a shark."

A second later both Meri and Cailey were holding their paddles over their heads and shrieking.

"Stop screaming," I shouted, adding to the din. "It's not going to attack us. Put your paddles in the water and start paddling."

They took off so fast I could almost see a rooster tail rising from the back of their kayak. I stifled a laugh. Then the shark began circling closer and cruised by just 6 feet away from my kayak, or at least it seemed that close. I could see that it was a salmon shark, her deep black eye looking right at me. It was a female, I knew, because only females come to Kodiak in order to feed on salmon returning to the streams to spawn. The males all stayed down in Washington or California somewhere, hanging around an offshore bar, the lazy bums. Salmon sharks came every summer and were commonly seen in Monashka Bay, near where we started our journey, because a small stream that empties into it has a strong run of pink and silver salmon. There weren't many salmon in the middle of the channel though, so the shark must have been attracted by the noise of our paddles. Deciding that I wasn't on her menu, she soon disappeared.

Meri and Cailey had entered the maze of little islets protecting Icon Bay from the ocean swells and were hunkered down in a cove, waiting for me. Happy to see that I was still in one piece, they paddled up to me as we excitedly discussed our amazing encounter for a few minutes before continuing our journey, meandering among the islets and coves until we reached the beach in Monk's Lagoon. There, we set up camp, made dinner, and built a fire on the beach to roast marshmallows before climbing into our sleeping bags for a well-earned rest.

The land around Monk's Lagoon is owned by the Ouzinkie Native Corporation and is part of the village of Ouzinkie, about 5 miles up the channel from Icon Bay. The word *Ouzinkie* is Russian for "narrow," because the channel is narrow, and the village sits at its narrowest point. Although there were several no-trespassing signs around, we had visited Monk's Lagoon before; it was a common destination for adventurous sightseers, so we didn't think anyone would mind us camping there as long as we were respectful of the property.

That night, Meri and I both had disturbing dreams. I dreamed that some monks and Natives came to hold a church service, and we were camping right in the middle of their church. Meri dreamed that some local people came to kick us out. Still in the clutches of my dream, I heard a high-pitched droning sound, like that of an outboard engine. A minute later, I realized it *was* an outboard engine. I jumped out of my sleeping bag, threw on some clothes, and climbed out of the tent. Soon a small skiff pulled up on the beach, and out climbed a local Native and a man dressed like a monk. The latter we had met before; he was part of a group of self-styled monks who had settled in Monk's Lagoon several years previously. I say self-styled because they were not part of the Alaskan Diocese of the Russian Orthodox Church, to which all the local Natives belonged. Most of the new monks were from California, and we had become friends with several of them. The new monks had built a very nice church or monastery in the woods just behind the beach and had begun to live there. But they had done so without permission from the Ouzinkie Native Corporation and eventually were told to leave. Nonetheless, they had remained on speaking terms with the Natives.

On this day, one of the Natives and one of the monks came to examine the church of Saint Herman, a mile back in the woods. They quizzed us briefly. Did we know what this place was? That it was holy to them? That it was private property? Yes, we answered, of course we knew, and that's why we had come to see it. We assured them we would respect the place and leave no trace of our visit, and that seemed to reassure them. After all, we were not the first visitors, and more would surely come. In fact, their visit was in preparation for the pilgrimage that would take place in August, a few weeks later. Expecting the arrival of up to a hundred people, they were preparing the church to receive the visitors.

Father (now Saint) Herman had lived and died in a small hut in the woods, several hundred yards up from the beach. One hundred years after his death, in the 1930s, a new church was built farther back in the woods, and Saint Herman's body was buried beneath it. That church was now over seventy years old and was starting to decay. Its foundation was rotting due to its location in the middle of a rainforest. Because of the condition of the church, Saint Herman's remains had been moved to Kodiak several years earlier and reburied under a replica of his original church. Over the summer, carpenters and volunteers had come to work on the church in the forest at Monk's Lagoon. Slowly, they replaced the foundation, the roof, and the siding, and even built a large deck that was able to hold the crowd that would come for the pilgrimage.

After breakfast, we followed the path back into the woods toward the church. Every few yards, we found small wooden plaques with a painting of a saint nailed to a tree by the path. During the pilgrimage, the faithful would stop at each icon for a short prayer. Today we just looked and remarked at the beautiful setting, surrounded by stately spruce trees and enveloped with wet, dripping moss. It was a celebration of green, punctuated periodically by bright red salmonberries. We dallied along the path, picking and eating the ripe, juicy berries until we had our fill. Before reaching the church, we stopped at a spring. Supposedly, Saint Herman had come to drink his water here, and a small shrine now stood at the site, with a metal cup for thirsty travelers to use. We all took a drink, honoring the site.

A small wooden hut and two graves stood nearby. One of the graves was for a man named Father Gerasim, an Orthodox monk who had lived there for a number of years in the early twentieth century. The other belonged to Father Peter Kreta. Father Peter had recently been the priest of the Russian Orthodox Church in Kodiak, following in the footsteps of his father before him, and was much beloved by the church members because he had grown up in Kodiak. A few years ago, he had been stricken with cancer and had died just last year in his early forties, leaving his wife and two young sons. His last request was to be buried near Saint Herman's church, on Spruce Island. Meri and I had personally known Father Peter and his family, and we were saddened by his death. We paused a few minutes to remember him and then returned to our campsite.

From the beach where we camped, we could look out and see a string of small islands jutting out from the south side of the bay. Most of these would have been here when Arkhimandritov visited in 1860. His journal indicated that he stood on one of them when he took a bearing to the *Kad'yak*. When I looked at the chart, it seemed that he could have been writing about the easternmost one, the third island. But I could not see it from the beach. In fact, it looked like just a pile of rocks, barely visible at low tide and almost completely

Monk's Lagoon, as seen from the shoreline during the kayak trip in 2002.

submerged at high tide. I realized then that he must have been standing on an islet to the west of it, the second island from shore, that was slightly larger and had grass growing on it. It was right next to the first, larger islet that rose up from the water in a steep cliff, about fifty feet high. During the great earthquake of 1964, this part of Kodiak sank up to 6 feet, but since that time, it has rebounded slightly every time an earthquake occurs (such as the 7.0 temblor that had occurred in January of 2002). Was it possible that the third, smallest islet had been above water when Arkhimandritov visited it? Or that the second island had broken off from the first during the great earthquake? What other changes had occurred?

That afternoon, we got into our kayaks and paddled around to the north side of Spruce Island, around East Point, and through a narrow, shallow channel over some rocky reefs. We pulled into a cove to have lunch on the beach. We had planned to go farther, but the weather forecast was starting to worry me. It was calling for 15- to 25-knot winds that evening and overnight with rain. That was not a good sign, so we decided to head back to camp. We came around East Point directly into the wind, which was now blowing about 20 knots—too windy for kayaking with a twelve-year-old. The quickest way back was to go through the channel out into the middle of the bay and across it, but that would expose us to the full force of the wind. Worse yet, the wind would be at our backs, which was dangerous because we wouldn't see the waves coming at us and could easily be turned sideways and rolled over.

After weighing the options, I decided on a slightly longer but safer course, following the shoreline inside of the kelp beds and rocky reefs. At one point we had to pass through a narrow break in the rocks, where swells were washing through. Timing our passage carefully, we waited for a swell to pass. Then we paddled rapidly into the cleft; as we did so, the next swell lifted us up and pushed us through. It was exhilarating but scary, like surfing. In truth it wasn't that dangerous, but having my daughter along on this trip made everything much more terrifying. I could expose myself to certain calculated risks, but I was not willing to do it with her along. That evening the rain started, so we retreated to our tent early. All night the wind blew and rain beat on the tent, and I found it difficult to sleep, finally dozing off in the wee hours.

That night I had the strangest dream. I was walking on a beach, with the ocean on my left and a forest to my right. Ahead of me, I could see something like a large skeleton, maybe the ribs of a whale, sticking up out of the sand. I walked toward them for a while, but they didn't seem to get any closer. Then an old woman came out of the woods. She was dressed in a nondescript sort of tunic and wore a shawl over her head. Was she young or old? I couldn't tell. She walked up to me and stood between me and the object of my interest, whatever it was. Then she pointed at me and began talking in a language I did not understand. I listened for a minute, uncomprehending, and finally I walked past her, only to discover that the structure had vanished. Where did it go and what had it been? Was it a whale skeleton? Or maybe the ribs of a ship? Who was the woman? I turned around to look, but she was gone. I woke up in wonder. What did it mean?

Daylight comes early in the Alaskan summer, so by six I was up and making breakfast. We ate quickly and packed up our camp, stuffing the wet tent and other items into the kayaks. It had stopped raining and the wind had come down, but the waves crashing on the beach still troubled me. Half an hour later we were out in the channel, and the water was surprisingly calm. The storm had mostly blown itself out, and we had to paddle through some small chop, but the swells had lain down. After a while, I began to relax. The rest of the trip was delightful, and several hours later we paddled up onto the beach behind Miller Point, tired but relieved that our journey was finished.

Afterwards, I thought about our visit to Monk's Lagoon. I wasn't a religious person; I didn't attend church, and I certainly didn't have any leanings toward Russian Orthodoxy. But I believed that I was a spiritual person, and that being so didn't require me to be religious. If anything, I worshiped nature. The power, the beauty, and the inspiration I find in nature seemed to be the same thing most people find in God. And Monk's Lagoon was a spiritual place. For some, it was because of Saint Herman and what he did there. For me, it was because of the beauty of the surroundings and the incredible experiences I have had there. What it meant to Arkhimandritov, I cannot guess, but he probably saw it as a reminder of a bad experience. Nonetheless, having been there, I felt more connected to it. It gave me dreams. It spoke to me. It told me the *Kad'yak* was there and that I had to find it. But not this year.

The shoreline in Monk's Lagoon (Icon Bay), as seen from the *Big Valley*. The church was built on Ouzinkie Native Corporation land by the "new" monks, who were later forced to leave.

CHAPTER 7

NEW
DIRECTIONS

SUMMER–WINTER 2002: ONE DAY IN the summer of 2002, Dave McMahan walked into my office at the Kodiak Fisheries Research Center (KFRC). Dave was the chief archaeologist for the State of Alaska Office of History and Archaeology and worked in Anchorage. His job primarily involved managing, documenting, and protecting archaeological resources on state lands, including submerged lands. He investigated historical sites, such as "the Castle" that Alexander Baranov had built as his home and headquarters in Sitka, and occasionally human bones that turned up when modern humans disturbed ancient (or not-so-ancient) gravesites. Dave had just completed his certification as a scuba diver and had developed an interest in marine archaeology. He knew Mike Yarborough, who had suggested that Dave should come see me.

"What do you know about the *Kad'yak*?" he asked.

"Oh, a little bit," I teased. "Let me show you." It was like taking the cork out of a champagne bottle. I opened my mapping program and showed Dave all the lines I had drawn on the computer. After many years of drawing with pencils on paper charts, I had finally graduated to using electronic charts. I could draw lines all over them, then just as easily erase them and start over. Somewhere in that morass of red lines lay the *Kad'yak*. Dave's eyes lit up. He knew

I was interested in the ship but he didn't know that I had done so much work on it. Over the next hour we talked about the *Kad'yak* and what it meant for the history of Alaska. There had been lots of shipwrecks during the period of Russian Colonialism, but none of them had ever been found. Most were poorly documented, and their locations were not well known. But the *Kad'yak* was well documented and still could possibly be found. If discovered, it would be the first ship from the Russian period ever to be located. Dave knew a number of other archaeologists around the country who he thought would be willing to help. He put me in contact with Dr. Tim Runyan, Director of the Maritime Heritage program at East Carolina University (ECU), in North Carolina. As soon as Dave walked out of my office, I was on the phone with Tim. That connection was the missing link in my quest.

Tim's specialty was investigating shipwrecks of the East Coast. His biggest accomplishment to date was locating the *Queen Anne's Revenge*, reputed to be the flagship of Blackbeard the Pirate. It had been scuttled in Beaufort Inlet, North Carolina, and Tim had been working on it for several years. But he was also interested in the *Kad'yak*. In fact, he had a Russian graduate student who wanted to do her thesis on the ships of the Russian-American Company. We immediately agreed to collaborate on a proposal to search for the ship. I would provide the local expertise, and Tim would provide the archeological credentials.

We put our heads together and came up with a plan. By early December, we wrote and submitted a pre-proposal to the NOAA Ocean Exploration Program (OEP). Now all we had to do was wait.

JANUARY 2003

On January 15, I received a letter from the Ocean Exploration Program saying that our pre-proposal had been approved and a full proposal was requested. I was ecstatic. I spent the next three days working frantically on the proposal, putting in all the details I could and poring over the budget details. When I had done as much as I could, I emailed it to Tim. I expected to submit the proposal through my agency (NMFS), and I needed the budget details for the ECU folks. Over the next few days, I checked my email hourly, expecting some response from Tim. It wasn't until Wednesday, January 22, that I finally got a message from him, saying that he was working on the proposal and would get back to me as soon as he could.

The next morning, I called Tim to ask about the proposal. We decided that it would be better for ECU to submit the proposal, with my agency as a subcontractor. That was easier for me because I would not have to fill out all the paperwork required to have it approved by the Alaska Fisheries Science Center (AFSC) in Seattle, a process that would add two or three days to the preparation time. Tim also said they wanted to beef up the methods descriptions. I was fine with that, though I thought it was fairly complete when I wrote it.[1]

TIME DRAGGED ON. I WENT in to my office at the KFRC over the weekend and checked my mail several times but found no messages. On Friday, I contacted Dave Kaplan, Director of the Baranov Museum, and Balika Haakanson, a teacher at Kodiak High School, and requested letters of support from them. They responded positively and provided us with very supportive letters of recommendation. I was very happy with that, especially with the opportunity to help Balika develop a lesson plan around the project for middle school science students. I also received a letter of support from Stefan Quinth, a Swedish filmmaker who wanted to videotape the search. Stefan was an internationally recognized nature photographer who had spent several years crashing through the brushy thickets and salmon streams of Kodiak Island to film the great Kodiak bears, and he was well known around Kodiak as somebody who could film just about anything.

The day before the full proposal was due to be sent out, I finally got an email from Tim. He wanted to keep the budget under $90,000 and wondered if we could cut the boat schedule from fourteen to eleven days and include more salary for one of his employees. I originally had only wanted six days of boat time. Eleven should be plenty, but I felt like the budget was getting whittled away with salaries. What were all these people going to do? The real work was in the logistics, diving, and on-site supervision, and I was reluctant to see too much of the budget go into salaries. We needed boat time and equipment. I needed a new drysuit. At this rate, I didn't think

I'd get to see Tim's budget until it was too late to change anything.

A week went by, during which I checked my email anxiously, waiting for a copy of the proposal. Nothing. Finally, I received a notice from NOAA OE that they had received the proposal. But where was my copy? Another week went by. Again, I called Tim to ask him about it. He had been too busy to send me a copy, but he'd get it out as soon as possible. A few days later, it showed up. As it turned out, it wasn't radically different from my original version. The ECU folks had added in costs for leasing their equipment, the magnetometer, and side-scan sonar—the latter of which I though would be useless in Icon Bay—plus their overhead costs. The final budget was a whopping $136,000. This seemed crazy. I supposed it really cost that much to do what we said, with travel and salary for all those people. But I really was beginning to wonder: Was all this necessary just to go search for a ship? After agonizing over it for several weeks, I finally decided just to let it go and hoped that the proposal would get funded.[2]

MARCH 2003

Over the past two months I was extremely busy in the lab, since I didn't have a field research project this spring. Nonetheless, I managed to let other things get in the way, like travel. I got invited by the Canadian Department of Fisheries and Oceans to help review their stock assessment procedures for snow crabs, and for some reason I agreed. Two weeks later, I was flying off to frozen Newfoundland. St. John's was cold, windy, and bleak, but I enjoyed spending time there. As I got used to the place I began to realize how similar it was to Ireland. The little pubs, fish & chip shops, and tea shops were so similar. If I sat and listened to the brogue, I could easily imagine myself in a pub in Galway.

A week after returning from Newfoundland, I learned that I had been awarded a year of funding by the North Pacific Research Board to study blue king crabs. I was surprised and elated. I would have enough funding to buy a new microscope, hire a full-time technician, and travel to scientific meetings. But that same week, I also heard that the Kad'yak project did not get funded. My high was brought to a low. To say I was greatly disappointed would be an understatement, and I couldn't bring myself to read the reviews and comments on the project. But I didn't have time to think about it or talk to Tim about it because I had to leave town again.

APRIL 2003

Early this month, I took a bunch of middle school students in a hand drumming group called the Kodiak Island Drummers, or KID, to perform at the Camai Fest, a big Alaska Native Dance Festival. KID was started by Michael Daquioag, who was the director and teacher. Having been a drummer (professional and otherwise) since the age of thirteen, I offered to help co-teach the group. The group included about fifteen students and a few adults, and we played on African drums called djembes and djun-djuns, as well as congas, bongos, and tube-shaped drums called tubanos. I wrote and taught some of the arrangements. Most of the time we played in Kodiak, but every year we took the kids on a trip somewhere in Alaska with funds we raised from our concerts.

Camai Fest was held in Bethel, way out in Western Alaska, though it took only an hour by jet from Anchorage. People came in from villages a hundred miles away on snow machines, traveling down frozen rivers or across the tundra for the festival. KID performed three times for a half hour each, a very generous performance schedule, especially compared to our fifteen-minute set that we were allowed at the Anchorage Folk Fest last year. There were many Native dance performances at the festival, and the more I watched, the more I was able to see the differences in costumes and dance styles from all the different villages. It was a wonderful trip and helped me forget about work for a while.

But I was back to crabs the following week when I traveled to New Orleans for a meeting of the National Shellfisheries Association. I thought I would have an incredible time enjoying the music there, but after a day and a half wandering around in the French Quarter without finding much in the way of real jazz, I was ready to go home. Eventually I found my way to Snug Harbor Bistro, outside the tourist zone, where I sat in the front row listening to Charles Neville wail away on his sax with a quintet in a tiny historic bar. The drummer was amazing, barely eighteen years old. I sat back and sighed. This was what I came for.

ALMOST AS SOON AS ARRIVING in NOLA, as they call it, I had a sore throat, and soon I was coughing constantly. Even though I had planned to stay an extra day past the meeting, I went home a day early instead. As it turned out, several of our adult drummers also caught the same coughing crud during our trip to Bethel.

I was sick for over a month with the cough. It just wouldn't go away. I couldn't do any scuba diving, and I was still depressed about the rejection of our proposal. I was in a complete funk. Eventually I resigned myself to the fact that we would not be searching for the *Kad'yak* this year. Only then did I finally get enough courage to read the reviews of the proposal. It must have been pretty bad to be rejected, especially with Tim Runyan as the project leader. In scientific research projects, the lead researcher is called the principal investigator, or PI for short. Tim would be our PI.

The proposal reviews included a number of minor comments, but two in particular seemed to be the source of rejection. All my worry about budget inflation were for naught because one reviewer concluded that we *had not asked for enough money* for various aspects of the project, particularly the public outreach section and materials to develop a school curriculum for teachers. The other reviewer's comments were particularly stinging; he did not think this project was of national significance. Never mind the difficulties of the search, the unknowns, uncertainties, and general crapshoot nature of the project. *It wasn't of national significance.* In other words, even if we did find the ship, it wasn't important enough to the nation to spend research funds on it. Imagine. I was steamed.

One aspect of the rejection took on a positive side, though. I realized I wouldn't have any major obligations toward research projects this fall. Usually at that time of year, all of our crab projects go into maintenance mode. The molting and mating season would be over, my experiments with juvenile crabs would be wrapped up, and then the fall would be spent mostly just keeping things alive while I analyzed data and worked on manuscripts and publishing. With the crab work slowed down and no funding to search for the *Kad'yak*, that meant I could go do something I had thought about for a long time. It was time to think about going to work on a temporary assignment somewhere else for a while.

The somewhere I had in mind was NOAA headquarters, and the particular assignment was something I wanted to develop at the National Undersea Research Program. Besides, my boss, Bob Otto, was dropping hints that he wanted to retire or resign his position as lab director. If that happened, things would certainly change for me. Either I would be hired to replace him and have a lot more responsibilities as the new director, or someone else would be hired for the job and would be putting a lot of Bob's responsibilities on me. If I was going to leave town for a while to pad my resume, this was the time to do it.

Once again, plans to search for the *Kad'yak* were put on the back burner. But not for long.

CHAPTER 8

SERENDIPITY
AND
REVELATION

MAY 2003: ONE DAY I stopped at the local video store to return a movie and ran into Josh Lewis, my daughter's sixth grade science teacher. Standing on the street corner, we started talking about diving. He was a diver and had been doing some exploration of recent wrecks on the west side of Kodiak with a friend of his named Steve Lloyd, who owned a bookstore in Anchorage. When I told him about my interest in the *Kad'yak*, he offered the use of his boat and several volunteer divers to go search for it. Together, standing on the sidewalk, we hatched a plan.

My spirits were high now that I had a diving schedule to look forward to. At last, I would get the chance to look for the *Kad'yak*.

JUNE 2003

I spent the first part of this month visiting NOAA headquarters in Silver Spring, MD, to arrange my temporary assignment there, and look for housing. Over the past month, Josh and I had grown more excited about the *Kad'yak* as we developed our plans. We were going to visit Icon Bay and dive over three days, July 20–22. In order not to jeopardize my job, I would take those days off and do the work on my vacation time. Besides myself, Josh, and Steve, the team would include a select group of divers. At the top of the list were Mark Blakeslee and Bill Donaldson, and local dive shop owner Verlin Pherson. We also invited Stefan Quinth; he was also a scuba diver and wanted to document the discovery process, if we found the ship. Since this was a volunteer operation, we would have to rent tanks from Verlin as opposed to using NOAA gear. To say I was greatly excited about the prospects would be the understatement of the year.

Underlying all the excitement, a dilemma bubbled to the surface. I was a scientist, well known in Alaska and especially in Kodiak. If I undertook an expedition to find the *Kad'yak*, it had to be completely above suspicion. I couldn't have anyone think that I was doing it for personal profit or aggrandizement. I had tried to do it the right way for years, teaming up with professional archaeologists and writing proposals, but nothing had come of it. Here was an opportunity to do it on the cheap. No proposal, no budget, nobody looking over my shoulder. Just hop on the boat with Josh and go do it. It was simple, direct, and appealing. But it was also wrong. Even if we found the *Kad'yak*, I would be a target for criticism. There was only one way to make this work, and that was to make sure that Dave McMahan was involved and the search was sanctioned by the state. Direct participation by the state archaeologist would ensure the state that any discoveries would be handled responsibly and within the law, and would negate the need for a permitting process to involve another credentialed archaeologist. At the first opportunity, I called and invited him to join us for the trip. He said he wouldn't miss it for the world.

With the date for our trip to Icon Bay rapidly approaching, I still did not have a good location for the *Kad'yak*. In preparation for the trip, I spent much of my time replotting all of Arkhimandritov's bearings. I had done this so many times before that I lost count, but it was always on paper. Now I was using a computerized navigation program that made it a lot easier, because if I was not satisfied with the lines and points, I could just move them around.

THE BEARINGS RECORDED BY THE captain were difficult to interpret for three reasons. First, he did not make them using 360-degree compass notations but rather quadrants. One of his first bearings was recorded from the "south point of Ostrov Elevoi," or Spruce Island, and was "5 degrees SW" to the right promontory of Monashka Bay. Fortunately, that location was still labeled "South Point" on the NOAA navigation chart that I was using, so I had a good estimate of his starting position. But the bearing could have four different meanings: It could mean five degrees west of South (185 degrees), five degrees south of West (265 degrees), or five degrees from Southwest (220 to 230 degrees). Over several months, I had consulted sailing manuals, *Bowditch* (Nathaniel Bowditch's navigation manual), and every historical document I could find regarding sailing, but I never came across any similar manner of recording bearings. Finally I decided that South would be the most prominent direction on the compass, and the bearing should be west of that, or 185 degrees.

Second, I assumed that Arkhimandritov's recorded bearings were magnetic rather than true bearings. Mariners have always used compasses to determine their direction, but compasses are magnetic and thus susceptible to variations in the earth's magnetic

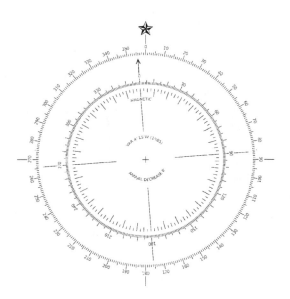

Compass rose showing negative (western) magnetic variation of -4 degrees, 15 minutes (as in Chicago, Illinois). My interpretation of "5 degrees SW" was 185 degrees magnetic, which in this example would become 181 degrees after adding magnetic variation.

field. Depending on your location on earth, a compass may point anywhere from a few degrees to tens of degrees away from true north. This difference, referred to as magnetic variation or declination, changed annually. The magnetic variation near Kodiak was now 19.5 degrees, but what was it in 1860, when Arkhimandritov took these bearings? Although compasses can be adjusted to read true north, it is unlikely that Arkhmandritov had such a compass with him as he surveyed the coastline of Spruce Island in a kayak.

Third, I assumed that if I applied some amount of magnetic variation to the bearings recorded by Arkhimandritov, I would get the true bearing.

These were all reasonable but major assumptions. How could I know what was right? The captain had recorded thirty bearings from thirty different landmarks before he wrote, "The mast of the *Kad'yak* lies on this course." Following that trail was like trying to follow a trail of bread crumbs through the forest after the squirrels have spread the crumbs all over the place. Which way should I go?

The last time I did this, in 2002, I used an estimated magnetic variation of 20 degrees east. When I did that, the first bearing, to Monashka Bay, was about 2 degrees too far to the east. The second bearing was supposed to be a line to Mys Melnichnoi, or Miller Point, located at Fort Abercrombie; but instead, it was almost a direct line to Spruce Cape, about 3 miles away and 10 degrees to the east. Was this Arkhimandritov's mistake or mine? These two bearings, shot from the same location, have a constant angle between them. By lining them up with the correct landmarks, I should be able to determine the correct magnetic variation. Moving them both 2 degrees west put them both dead on to both landmarks, but one still directed me to Spruce Cape rather than Miller Point. Based on this experiment, I concluded the correct magnetic variation was 22 degrees east, so that the true bearing would be 185 + 22, or 207 degrees. But the Spruce Cape–Miller Point confusion bothered me. If I couldn't interpret the first bearing correctly, then I couldn't trust my interpretation of any of the other bearings either. Perhaps sitting low to the water in a baidarka, rocking in the waves, and trying to take bearings with a compass and a sextant, the captain could not tell what point he was looking at. Possible, but not likely.

ON JUNE 30, I WENT to meet with Stacey Becklund, who had replaced Dave Kaplan as director of the Baranov Museum. Dave had suggested the museum might have some money they could use to help our project, so I planned to ask for only $1,000 to help defray the costs of fuel and renting some dive equipment. Hearing this, Stacey invited me to present my case to the board of directors for the Kodiak Historical Society (KHS), who ran the museum. They were scheduled to meet a week before our planned dives. Since I was already at the museum, I spent some time examining their book collection. Of particular interest was a reference in one book to Mikhail Tebenkov's *Atlas of the Northwest Coasts of America*, a series of Alaskan maps published in 1852. There was also a dictionary of Alaskan place names. I checked to see if Miller Point had ever been used to refer to another place, but it only listed the current location at Fort Abercrombie.

After the museum, I went to see Patrick Saltonstall, an archaeologist at the Alutiiq Museum. I thought maybe he could help me locate the "old and new villages," which were some of the landmarks referred to by Arkhimandritov. Patrick only knew of one potential village site, called Perekovsky Beach. Apparently, it was occupied by an old Russian man with one or more Native wives and multiple children. If I really wanted to know about Spruce Island, he suggested, I needed to contact Lydia Black.

"Lydia knows more about the Russian history of Kodiak than anybody alive," he said. "She is completely adamant about her knowledge. However, it may not be correct—but don't tell her that."

Lydia Black was a retired history professor at the University of Alaska and an icon among Alaskan historians. Revered for her knowledge about Russian America, she probably knew more about that topic than any person alive. She was born and grew up in Russia under Stalin, and spent time in a German labor camp during WWII before coming to the United States, where she earned her PhD in anthropology. As a result, she has read and translated many old Russian documents about life in Alaska and was in the process of writing a book about it. She also happened to live in Kodiak. I had met her once before and mentioned my interest in the *Kad'yak* briefly, but she seemed to dismiss it as though finding the wreck was unimportant and I were just a dilettante. For that reason, I hadn't talked with her about it again. Perhaps it was because of her reputation, scholarly but dogmatic. Perhaps I didn't think she could help. Perhaps I just didn't want to be rejected again. But this time I knew I needed help, so putting pride aside, I called her on the phone.

When I told her I wanted to know about the villages Arkhimandritov referred to, she was immediately curious and wanted to know why I was interested. I told her I was looking for the *Kad'yak*.

"It didn't sink," she replied gruffly. "They salvaged it."

I was stunned. Was this the end of my search? Was I doing all this work only to learn that there was no ship? I stammered a few times and then protested. "But Arkhimandritov surveyed it," I said. "I have his notes, translated from Russian."

That really got her interest. She wanted to know where I got them and who translated them. What I told her seemed to pass muster, since she knew the translator personally.

"Oh yes," she said, as memory slowly returned. "That's right. It did sink."

I felt relieved, but couldn't believe I was being yanked up and down like a yo-yo. Then she paused, as if trying to decide whether or not I was serious. "You should come see me," she said.

A minute ago, she had sunk me into a bottomless pit, then just as suddenly she pulled me out. She could change her mind if given the right information.

Fifteen minutes later, I was knocking on Lydia's door. A short, elderly woman answered the door, slightly stooped over, her gray hair wispily going in many directions. She invited me into a tiny apartment, and we sat at her kitchen table. I showed her my materials—the translations, the diagrams, and my maps.

"Oh, yes, she said, "this was translated by Kathy Arndt."

When I told her about the Miller Point–Spruce Cape mix-up, she was adamant. "Arkhimandritov wouldn't make such a mistake. He grew up here, you know. He lived here all his life. Let me show you."

In the middle of her small kitchen was a table covered with papers, stacked at least a foot high. From the middle of the pile she deftly extracted a manila envelope, careful not to upset the pile, like a stack of Jenga blocks. Opening it carefully, she then pulled out a map. Lydia had been to visit the Russian Naval Archives in St. Petersburg the previous year and had photocopied an original chart of Kodiak drawn by Arkhimandritov and dated 1848.

"See there," she said, "that is Miller Point, known then as Mys Melnichnoi, or 'Mill Cape.'"

From previous experience (both working on Russian ships and studying Russian language), I could read the Russian script well enough, but to my surprise, Arkhimandritov had placed that label on what we now called Spruce Cape. Suddenly, something clicked in my brain. Goosebumps rose on my skin. If I were in a cartoon, then a lightbulb would've appeared above my head.

"Look," I said, pointing. "What he called Miller Point is really Spruce Cape."

She agreed, mostly that he knew about Miller Point, but I'm not sure she really understood the significance of my revelation. I had drawn his bearing as he recorded it, plus my estimate of

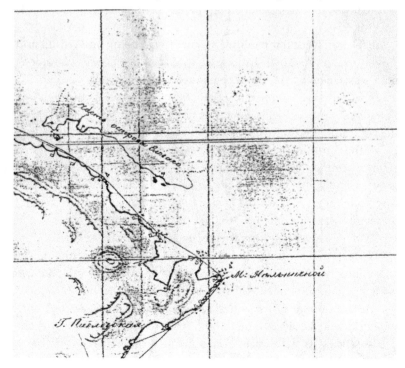

Left: Original map of NE Kodiak Island made by I. Arkhimandritov in 1849 (*From Russian Naval Archives, provided by Lydia Black*). "M. Melnichniyi" (Mill Cape) is notated at rightmost point (now known as Spruce Cape).
Above: The version of this map published as part of the Tebenkov *Atlas* in 1852; "M. Melnichniyi" has been relocated to the current location of Miller Point, and replaced by "M. Elevoi" (Spruce Cape), as it has been known ever since.

magnetic variation, and it pointed to what we now called Spruce Cape, but which Arkhimandritov knew as Miller Point. The name "Miller Point" has been moved since then and was now associated with Fort Abercrombie. That meant that my interpretation of his bearings was correct.

She nodded, perhaps understanding that the names had changed, but not at my enlightenment. Lydia told me that Arkhimandritov had made a complete survey of the coast in 1860, from Kodiak to Afognak Island. It had taken him six weeks. He had traveled in two three-person baidarkas with two paddlers. His first stop, she said, had been at Perekovsky Beach before traveling along the south side of Spruce Island and then north.

"But," I protested, "his notes say he started at the south end of the island, what I think is South Point."

"No, he really started at the northeast end of the island, at a settlement of hunters," she said.

"Would that be Selenie Bay?" I asked.

"No, Selenie Bay is on the south side, near Ouzinkie."

That didn't help. I needed to find a Selenie Bay near Monk's Lagoon in order to make sense of Arkhimandritov's bearings.

"What does *Selenie* mean?" I asked. She told me it meant "settlers" or "settlement." Again we looked at the map. Written very small, close to Icon Bay, was the word *Selenie*.

"There," I said, "is that Selenie Bay?"

"Maybe, but Selenie could refer to any settlement or group of settlers," Lydia replied.

But there it was, written just where I thought it should be, in what we now called Icon Bay, or Monk's Lagoon.

I tried to tell her about my plotting all the bearings and that they did not point to the ship directly. She only seemed to know about three bearings that were the last ones Arkhimandritov took. They were probably shot from Ostrof Point because they all lined up to that spot. She finally agreed with me that he must have started his survey from South Point, but she insisted that he traveled west along the south shore and did not survey Icon Bay at all.

I don't think she understood how meticulously I had plotted these bearings. She seemed to know a lot about his overall journey, which

I did not, but she didn't seem particular about the details. For one thing, she said, "The right and left promontories of Monashka Bay are the bluffs as seen by a ship leaving the bay." But Arkhimandritov's notes only made sense if they were the right and left sides seen while entering a bay. A mariner would need to know how to identify a bay from out at sea; it wouldn't help him once he was in the bay.

Before I left, she gave me a roll of microfilm and offered to help interpret it. I said I would have a look at it and then call her.

After leaving, I had a very unsettled feeling. Was I all wrong about Arkhimandritov surveying Icon Bay? Did I mislead myself into following his bearings in that direction because I presupposed that's where the ship was? Could it actually be farther up the channel to the west, toward the village of Ouzinkie?

I sat down and reviewed my interpretation of the bearings recorded by Arkhimandritov and translated by Kathy Arndt (see Appendix A). Some of them lined up perfectly with known landmarks; others were totally uninterpretable. Arkhimandritov had written he saw an island in this direction, but from what I knew that would mean he would have been looking into the woods—unless I moved his position over a few hundred yards. Then the bearing worked. In fact, all the other bearings fell into place right after this one. Was he shooting them from a kayak? If so, the kayak may have drifted or been paddled between the time he took three different bearings. Could I possibly be wrong?

Looking at the list of bearings, all of the first ten were northeast. Even using a quadrant system, I couldn't get that wrong. Thirty-five northeast was 35 degrees plus magnetic variation, whatever it was, but it was still northeast anyway you plot it. Northeast was northeast, however you write it. He had to have been traveling northeast, from South Point, into Icon Bay. How could this possibly be interpreted any other way? If he were going northwest up the channel to Ouzinkie, the bearings would all be 90 degrees to the west.

Even given the few uninterpretable bearings, they led me to the same point. I saw now that a few of these bearings connected. He must have gone in this direction, then that, then another, and suddenly he was there, at a place where many bearings lined up. There was no other way to read it.

Another thing that puzzled me was why, in all these notes, he had taken only one bearing on the mast of the *Kad'yak*. Why hadn't he taken three, from different directions? That's how a good surveyor would have done it. I guessed perhaps because he could see the mast. Its location was no mystery. There it was, sticking out of the water. Nor had he noted the distance from his location. It hadn't been the focus of his survey— just a miscellaneous side note, a marker in the water. A nagging reminder of an unfortunate event in his life. There it was, he wrote. It needed no further remarks.

If I was correct about the location from which he took the bearing, then the *Kad'yak* had to be about halfway between him and the opposite shoreline of Icon Bay. There were rocky reefs jutting out about 100 yards. Their depths were noted on my computer, recorded from the visit there with the ROV in 2002. The ship would have to be in the middle of a narrow channel, about 70 feet deep. Right in the middle of the "c" in the words "Icon Bay" on the map. That was my bullseye.

Three times I had plotted these bearings. Three times they had led me to the center of Icon Bay, though the final position moved a few meters. Herein was the puzzle to be solved. Arkhimandritov did not record the exact position of the ship for posterity to find it. Perhaps he hadn't been interested in the exact position. What he did leave us was just a trail of bread crumbs. Thirty of them, from his starting point to the ship. It was up to me to connect the dots and follow the trail. If I did it right, it would lead me to the ship.

WEDNESDAY, JULY 2, 2003

Today I dug out the original materials about the wreck that I had received from Mike Yarborough in 1991 and read through them all, just to see if I had missed anything. There, I found an envelope containing Tebenkov's *Atlas*. Mikhail Tebenkov was a captain in the Russian Navy and had served as the manager of the RAC from 1845 to 1850, when Arkhimandritov surveyed Spruce Island. Tebenkov was the first person to create an atlas of the Alaskan coastline by assembling charts drawn by all the Russian ship captains who had sailed there. I had purchased the atlas a few years ago at the Alaska Sea Life Center in Seward thinking that it might come in handy someday. But I had completely forgotten about it, or perhaps didn't realize what it was.

The atlas consisted of a folio of forty large maps, folded over and stuffed into a manila envelope. Each map covered a section of the coastline around Kodiak and other nearby islands. Included in the package was a full-page chart of Kodiak. To my surprise, it showed Mys Melnichnoi, Miller Point, in its current location at Fort Abercrombie. The date on the map was 1849, but it had been published in 1852. Reading the documents with it, I learned that the map had actually been surveyed by Arkhimandritov. On his original map of 1848, he drew in Miller Point where he knew it, at Spruce Cape. Between that time and *Atlas's* publication four years later, Miller Point had been moved to refer to Fort Abercrombie. Who moved it? It must have been Captain Tebenkov. That was the origin of the mix-up. I sat back and breathed a sigh of relief. It verified my interpretation of the bearings, the location of Arkhimandritov's first sightings, and the 22-degree magnetic variation. Arkhimandritov must have known that location as Miller Point from his youth and labeled it as such when he first surveyed it. When he took the bearing from the wreck site to Spruce Cape in 1860, he still called it by the name he knew it as, Miller Point, despite the fact that it had been renamed in 1852.

That evening I went to the library and examined the microfilm Lydia had given me. It was titled (in English) "Journal of an overland expedition by Lt. Zagoskin, 1842–1844." The rest of it was all in cursive Russian. Not only was it too difficult to read, but it didn't seem to be relevant or from the right time period. I had thought she had given me Arkhimandritov's journal and made a note to go ask her about that. For now, one mystery remained: Did Arkhimandritov really travel up the south side of Spruce Island in a northwest direction, as Lydia insisted? Or did he travel northeast, into Icon Bay, as I concluded? My notes only included his journal entry from July 12, but not the next day. What did he do then? Did he turn around and go northwest after that? Seeing those notes would've helped me solve this problem. It would be difficult convincing Lydia that I was right. Perhaps I didn't need to have her believe me; but it would be a great boost if she did.

THURSDAY, JULY 3, 2003

I had hoped for another opportunity to paddle over to Icon Bay and spend a day re-shooting the bearings now that I had a good idea of what they meant, so I watched the weather daily. It was a weekend holiday, so I should have three days. Actually, tomorrow, Friday, I had to work a few hours in the day and play drums with KID downtown. So if I wanted to go to Icon Bay, I would have to go late at night then. The trip required a three-hour paddle across 5 miles of open water and a two-night campout. It was a major trip for me to undertake by myself. Meri, Cailey, and I had made the journey last year in somewhat iffy weather, 15 knots southeast most of the way over. We had endured two rainy nights with a strong 25-knot blow the second night that allowed little sleep as I worried about the return trip. But it had been an easy paddle back to town, with the wind at our backs and fairly calm seas.

The weather had been nice the past few days. Sunny, but very windy. The weather report called for high winds today, 20 knots northwest, that would die down this evening, but tomorrow was supposed to be cloudy with winds from the south. Then Saturday was going to have winds blowing 20 knots south or southeast—the worst direction because that blew right into Icon Bay. I sighed. I decided I had better not go. I would have to put the trip off for another week. I crossed my fingers for better weather.

Decision made, I loaded my kayak onto my car and headed down to the harbor. I joined a flotilla of likeminded kayakers, and together we paddled out into the channel to watch the fireworks being shot off from Near Island. The show began a few minutes after midnight, as July 3 rolled over into July 4.

SUNDAY, JULY 13, 2003

Weather over the holiday weekend had been as bad as predicted. It had rained all day Saturday and finally cleared on Sunday. That pattern had repeated itself last weekend too: rain on Saturday, sunny on Sunday. Paddling over there for three days wouldn't have left much time. Josh had also called last week to ask me if we could delay the dives by two or three days, to start on Tuesday or Wednesday instead of Sunday, which I was absolutely fine with.

And this coming weekend was the Bear Country Music Festival where I was planning on playing music every evening with my jazz and rock group, Phil Dirt and the Dozers, which meant Saturday would be a late night out. It wouldn't be good to dive the next day tired and sleep deprived.

But I still had some *Kad'yak* business to do. Today I met with the KHS Board of Directors to ask for a donation of $1,000 to help defray the costs of our search. My previous attempts to obtain larger sums of money from other granting agencies had been all denied for various reasons. I thought it boiled down to the fact that finding the *Kad'yak* was really a crapshoot until some hard evidence turned up, and the only way that would happen is if we go look for it. As we spoke, it was clear that the board was extremely interested in the proposal, especially because it would like the museum to be the repository of any artifacts that might be recovered. If such artifacts were preserved and displayed, they could form the centerpiece of a new annex, and it would help bring in many more donations and grants to help the museum. But the budget was too slim, the board said, to provide any support. The members also worried about the future ramifications of such an effort; what would it cost to preserve, curate, and display such materials? And how could they be assured the State of Alaska would not claim the materials for display at the State Historical Museum in Juneau? I could not provide them much assurance on any of these considerations. I could only remonstrate that I was committed to finding the *Kad'yak* and that we were going to look for it with or without their help. If they could find a way to help us out, it would be a great gesture toward laying claim to whatever artifacts might eventually turn up. But if not, we would bear the costs ourselves.

Adding to the board's concern was a move by Senator Frank Murkowski to place the Baranov Museum under control of the National Park Service. The museum was located in the Erskine House, one of the last remaining log constructions built during the Russian era. Originally built by Alexander Baranov as a fur warehouse, it was on the National Historic Register, but maintaining it was a strain on the museum's budget. The National Park Service could take care of all of that, but it would also take away some local control of the facility. None of this entered the conversation or discussion today though; it was only present as a faint echo in the background.

I left feeling satisfied that I had made my case. I had demonstrated my knowledge of the wreck and my commitment to doing the search in the most ethically responsible manner, and now everything else was out of my hands.

Later that evening, I received a call from Mary Monroe, one of the board members and my next-door neighbor. She told me that the museum couldn't make a donation from its funds, but individuals on the board would make personal contributions to our effort. The board had decided that it would like to support the search efforts in order to "get a toe in the door" of any future events concerning the wreck. Apparently, the board members had continued their discussion for three hours after I had left, and the decision they reached was to establish a research fund to which individuals could make tax-deductible contributions. It was to be the first research fund they have created, and all on account of my request. They all understood this was a pie-in-the-sky effort, Mary said, but they knew I was an ethical person and would do the best job I could. It was a pleasant surprise, and I thanked her for her support. I felt complimented that they would put their trust in me.

In another week we would begin our search for the *Kad'yak*. I was excited about finally having the opportunity to explore the bottom of Icon Bay by scuba diving. It would require me to bring all my training and experience to bear. As part of my mental preparation for that adventure, I had to consider not just the excitement and the fun of scuba diving, but the dangers as well.

CHAPTER 9

HAZARDS
OF THE
DEEP

For as long as humans have lived near the ocean they have wondered what lies beneath the surface. Sponge divers and other intrepid souls had been using heavy suits with helmets to explore the seafloor since 1865, but such suits were cumbersome and dangerous. Scientists and adventurers had long dreamed of a small convenient system to allow underwater exploration, one that would let them swim like a fish without being tied to either the surface or the bottom. They finally got their wish in 1943, thanks to two Frenchmen, Navy Lieutenant Jacques Cousteau and engineer Emile Gagnan. Cousteau and Gagnan adapted a demand-valve regulator used on natural gas tanks, which allowed them to breathe compressed air from a steel cylinder. They called their system "Self Contained Underwater Breathing Apparatus," or SCUBA, but the term is used so commonly now that it is rarely capitalized. Although modern scuba systems are far more complex, the basic concepts have not changed.

In order to breathe underwater, you must inhale air at ambient pressure, that is, the same pressure as the water surrounding your body. (Try sucking air through a garden hose from the bottom of a swimming pool and you'll see what I mean; on second thought, don't. There's a reason snorkels aren't any longer than a few inches.)

Air pressure at sea level is only 15 pounds per square inch (psi), but as you descend into the ocean, water pressure increases by 1 atmosphere (15 psi) every 33 feet, or 10 meters, so a diver at that depth is under twice as much pressure as when at the surface. When the diver inhales, the regulator delivers air at the same pressure as the water. As you go deeper, the air you breathe is more compressed, and you use more of it. A typical scuba tank holds 80 cubic feet of air compressed to as much as 3000 psi. That much air would last almost two hours at 30 feet, but only about 45 minutes at 60 feet. As long as the gear works properly, diving is safe. However, diving entails several inherent dangers.

The major hazards for divers are caused by nitrogen. Air consists of 21 percent oxygen and 78 percent nitrogen, plus some other gases, including carbon dioxide. Breathing compressed air causes the diver to absorb more nitrogen than at the surface, which can lead to nitrogen narcosis, or "rapture of the deep." The deeper you go, the stronger the effect, such that below 100 feet you feel like you have had too many martinis and can't think clearly. If you ascend to the surface too fast, nitrogen dissolved in the bloodstream can come out of solution, forming bubbles, like popping the cork on a champagne bottle. Those bubbles can cause severe pain if they get lodged in the joints, causing the bends, or what we now call Decompression Sickness (DCS). DCS can cause permanent damage, even kill you if bubbles get into your lungs or brain, impeding the absorption of oxygen.

As you ascend back toward the surface, air in your lungs expands. Ascend too quickly, and rapid air expansion can burst the alveolar sacs in your lungs, causing an air embolism, which can be fatal. Therefore, you must ascend slowly enough to allow both nitrogen and high-pressure air to escape your lungs. The default rule for this is to ascend no faster than 60 feet per minute, or 1 foot per second. A good way to gauge this is to watch your exhaled bubbles and ascend no faster than the smallest bubbles.

Another hazard is running out of air. That could require you to make an emergency ascent, swimming up to the surface with a lungful of compressed air, which could cause an embolism. Beginning divers are trained how to make emergency ascents safely by exhaling continuously during the ascent. They are also taught to always dive with a buddy, who can share their air, either by "buddy-breathing" or from a spare regulator. Modern divers carry a pressure gauge to monitor air pressure throughout the dive, so running out of air is less common.

After ascending, you still retain some residual nitrogen in your bloodstream, which can take several hours to completely clear, during which you must remain at the surface. We call this process off-gassing, or blowing off nitrogen. The US Navy has developed tables that tell divers how long they must off-gas before diving again, which depends on both the depth and the duration of the previous dive. These tables were developed by experimenting on young, healthy, physically fit males, so they don't necessarily reflect the condition or response of every diver, especially an overweight, middle-aged diver like myself. Diving within the depth and time limits described by the tables, or a computer, allows you to descend and ascend without additional decompression time. This no-decompression diving is the standard for recreational divers. Nowadays, though, every diver carries a computer, which constantly monitors your depth and dive time, tells you how much time you have left to dive, and how long you need to sit on the surface off-gassing before you can dive again. It can also tell you if you need to decompress, at what depth, and for how long. But every diver is different, so even following all the rules doesn't guarantee that you will not run into problems.

For that reason, you should always end the dive early enough to arrive at the surface with 500 psi left in your tank. That is enough to cover a few extra minutes of dive time, in case you run into trouble on the way up or have to share it with your buddy. As an additional safety measure, you should include a three-minute safety stop at 15 to 20 feet before surfacing, to allow additional off-gassing. Modern dive computers will tell you when to do this and count down the time for you. More experienced divers can spend more time underwater, often with two or three tanks, if they make multiple decompression stops at depths prescribed by their computers. The practical depth limit for scuba with compressed air is 130 feet—below that, divers become susceptible to nitrogen narcosis and oxygen toxicity. At 100 feet you only have about 15 minutes of bottom time, including the time spent descending and ascending, so you can't spend much time working or looking around. Longer bottom times can be achieved by replacing some of the nitrogen with additional oxygen, so-called Nitrox, or enriched-Air nitrogen, EAN32. In order to dive deeper, you need to use mixed gases in which the nitrogen is replaced with an inert gas such as argon or helium. But these gases are also more dangerous and require additional training to use.

Water temperature is another problem to overcome. Because of its density, water absorbs heat very rapidly from your body. Thus, most divers wear an insulated suit of some kind, even in the tropics, either a wetsuit or a drysuit. Wetsuits made of neoprene rubber allow a thin layer of water between the suit and your body that warms up quickly. Drysuits, on the other hand, keep you warm by keeping the water out and trapping air inside with tight rubber cuffs at the wrists, a tight collar, and a watertight zipper. However, most of the divers who use them know that there really isn't any such thing as a drysuit—they are all wet to some degree, due to leakage at the cuffs, neck, or zippers. Modern drysuits are thin shells of waterproof fabric, under which you wear a fleece or some other insulating material.

Both types of suits are buoyant and require weights to keep you down in the water. When I dive with a 7-mm wetsuit, I typically wear 18 to 20 lbs of weight (which is a bit heavy but keeps me from rolling around if there is any surge), whereas when I dive with a drysuit, I wear 35 to 40 lbs of lead shot, on a belt suspended from my shoulders.

The goal is to remain neutrally buoyant, neither sinking nor floating towards the surface. But that's tricky because as you descend, water pressure compresses both your body and your wetsuit or drysuit, causing you to become denser and thus heavier; the result is that you sink faster. To compensate for this, divers wear a buoyancy compensator, or BC, an inflatable vest connected to your scuba tank. Drysuit divers may not use a BC though because they can inflate the suit directly from the tank to control buoyancy.

When you enter the water, your body immediately initiates the amazing marine mammal diving reflex (MMDR). Humans, dolphins, and whales all share this response, which allows us to survive in a low-oxygen environment. As soon as water hits your face, your heart rate slows by 10 to 30 percent, and muscle tissue in peripheral blood vessels of your arms and legs contracts, forcing blood to your core body organs, including your heart, lungs, and brain. Slowly, your spleen releases blood cells into the bloodstream which increase the amount of oxygen that can be delivered to the core organs, the most sensitive to oxygen deprivation. Reduced blood flow in the limbs makes you feel cold but actually reduces heat loss to the external water. It also increases blood pressure in the kidneys, causing your bladder to fill up. It's no coincidence that getting in water can make you feel like you have to pee. This isn't a problem if you are a dolphin, but it is if you are a diver. As we say, there are two types of divers: those who pee in their wetsuits, and those who lie about it. If you are wearing a drysuit though, you just have to hold it.

To BECOME A SCUBA DIVER, you must undergo some training and acquire a dive certification, usually from a commercial organization such as PADI, the Professional Association of Scuba Divers, or NAUI, the National Association of Underwater Instructors. In addition to training and experience, divers who work for NOAA or for a university must adhere to rules and regulations established by those organizations. NOAA has its own set of regulations plus a thick diving manual, much of which was adapted from the US Navy diving program. Universities follow a set of protocols established by the American Academy of Underwater Sciences (AAUS).

I first learned to dive in 1965, when I was thirteen years old, in a rock quarry in Central Ohio, about as far from the ocean as you could be. When I joined NOAA in 1984, I had been diving for nineteen years, but they required me to attend the NOAA dive training program anyway. So, in 1987, I spent three weeks at NOAA dive school in Seattle learning how to dive in a drysuit. A few years later I spent another week training to be a NOAA divemaster, which allowed me to plan and supervise diving activities with other divers. By the time we began the search for the *Kad'yak*, I had made over 440 lifetime dives had acquired numerous dive certifications, including a Nitrox certification that I could use for recreational diving .

We take our modern equipment for granted, but in the early days of diving, not everybody had a pressure gauge. Scuba tanks of that era had a J-shaped valve that prevented you from breathing the last 500 psi of air. When breathing became difficult, you just pulled down on the valve with an extended rod, making the reserve air available for the ascent to the surface. The first scuba tank I owned had a J-valve on it that almost killed me before saving my life.

As a graduate student in 1978–80, I earned extra money by doing commercial diving around a marina in Westport, Washington. Over one hundred boats left the marina daily to fish for salmon, either recreationally or commercially. Often they would return to port with fishing line, nets, or other items wrapped around their propellers. When that happened, the skippers would call me and I would meet them at the dock with my scuba gear, dive under the boat, and remove the tangled lines from their prop shaft.

Once, I was working under a boat that had a large tangle of monofilament salmon gillnet wrapped up in the propeller. The spinning prop shaft had caused the net to melt around it, so I was using a hacksaw and a pair of metal shears to cut it free. The rest of the net was floating around me and was very difficult to see in the murky marina water. I removed the pull-rod from my J-valve and left it in the down, or open, position so that I could use all the air and monitor it with my pressure gauge. When it showed me that I was low on air, I put away my tools and prepared to return to the surface. Suddenly, I realized I could not move. I was stuck. Somehow, I had become entangled in the free-floating gillnet.

The first thing to do in such a situation, I knew, was *don't panic!* Take time to assess the situation and figure out a solution. Feeling all around, I realized that I was hung up somewhere behind me. I could feel the net behind my neck and tried to cut it with the shears, but I couldn't reach it. The hacksaw wasn't any help, and my dive knife wasn't sharp enough to cut the monofilament. With only minutes of air left, I knew the only way to extricate myself from this predicament was to take off my scuba gear by removing my BC. Careful not to become any more entangled in the net, I detached all the buckles and straps but still couldn't wiggle free. With a mighty jerk, I pushed against the boat bottom and kicked downward to free myself. Just then, I fell out of the net and sank to the bottom of the harbor, still wearing my BC and scuba tank. On reaching the surface, I realized that I had been stuck on the J-valve. The downward pointing valve had become hooked in the mesh of the net, and when I pushed myself away from the boat, the valve flipped up, freeing me from the net. I never used a tank with a J-valve again.

Scuba diving is a complicated business, and can be dangerous. But it can also be fun, exciting, and safe. And with that knowledge, we are now ready to pull on our drysuit, don our diving gear, and plunge into the ocean to begin our search for the *Kad'yak*.

CHAPTER 10

IN
SEARCH
OF THE
KAD'YAK

SATURDAY, JULY 19, 2003: ALL the dive participants arrived in town today. I spent the morning in the lab taking care of my crabs and just barely finished by the time we were all supposed to meet. I didn't even have time to go to the dive shop and pick up the tanks we would need for diving. After meeting Dave, Stefan, and Josh at Harborside, the local coffee shop along the waterfront, we headed out for the dive shop Scuba Do. There we met up with Bill Donaldson and Verlin Pherson, picked up a dozen tanks and weights, and loaded them into Josh's truck. Then we all went back down to the harbor to load our gear onto Josh's boat, the *Melmar*.

The weather report was not good. It was blowing 20 knots southeast today with gusts to 35, and similar conditions were expected for tomorrow. That would blow right into Monk's Lagoon and make a huge swell. After stowing our gear on the boat, we decided to reconvene at Harborside tomorrow morning at eight and make a decision then based on the weather.

I drove out to the fairgrounds for the Bear Country Music Fest and played with KID and Phil Dirt and the Dozers. But even while performing, I had a nagging feeling of worry because there were still some important items I needed to take care of. After playing my last set that evening, I drove back to town around nine. By that time, the wind had died, the clouds had partly cleared, and it was beginning to look like decent weather. I hoped it would hold for tomorrow.

Before going home, I drove over to the KFRC to pick up a portable GPS unit. The Global Positioning System was originally designed so that the US military could figure out where soldiers were. GPS has revolutionized the way we find our way around, and today, receivers are common in cars, phones, and even cameras. But in the early years of GPS development, the US Department of Defense, in its greater wisdom, decided it was too accurate for the average citizen and scrambled the signal so that commercial units would only have an accuracy of about 100 meters. But not long after GPS became available, engineers figured out how to overcome the scrambling. By establishing a base station at a known position, they could receive the GPS signal, correct the position, and broadcast the differences to other receivers; this system was called Differential GPS, or DGPS.

In 2003, our lab had three of the portable military GPS units, built before DGPS was widely available. These models were a dusky tan color and had literally been designed for tank commanders in Iraq. I located one at the lab and took it home, hoping I could get it to work. People always complain about the arcane rules of government contracting, and this device was Exhibit A. For some unfathomable reason, whoever put out the bids for this little baby stipulated that the units had to use batteries that were only made in France. This was either a perfect example of military intelligence or a nod to our French partners in the Iraq debacle. Whatever the reason, we could not purchase the batteries ourselves, so we had to send our units to Seattle every year to have them changed. They had just been returned to us, and the units needed to be reset to our current location. I should have done this earlier last week, but it had escaped my attention.

That evening, I sat outside on my deck until almost eleven, going through the menus, trying to figure out how to set everything in the system—date, time, location, position type, almanac definitions. I didn't have the manual, but I had done this before a few times and managed to figure it out. Finally, I turned it on and checked the position, but the last position it showed was in Seattle. It usually took a while to update from such a big difference. Turning on my laptop, I looked at the Kodiak chart, put the cursor about where my house should be, and checked the position. Then I keyed that into the GPS unit, took it back outside, and laid it on the deck. Half an hour later, I came back to check it, and it had finally updated to something close by. That was good enough, so I went to bed.

SUNDAY, JULY 20, 2003

Today was to be our first day of diving. I was already awake at six in the morning and could not go back to sleep, so I turned on the radio and listened for an hour, worrying about weather, diving, problems at the laboratory, anything I could think of. When I finally got up and looked outside, what I saw did not make me happy. My house sat on the mountainside looking out over the ocean and the mountains farther down the island. I could easily see that flags in the harbor were blowing straight to the west, good indications that there was a pretty strong easterly blowing because the wind makes a 180-degree turn as it comes around the island and sweeps through the harbor.

Josh said his boat could take the weather, but I had other concerns, the first being for everyone's safety. If we tried to dive in these conditions, someone could get hurt. Climbing in and out of a boat in 6-foot waves would be dangerous, and we would be sitting right in the swell all day. One of the primary rules for safe diving is this: If you can't ensure the safety of your divers, don't dive. And besides, the waves would stir up the bottom so much that the visibility would be crap anyway and we wouldn't be able to see anything. I also didn't want to waste everybody's time sitting on the beach or rocking around in a boat. What to do? I thought maybe we could delay four hours to see if the sea improved. I ate breakfast in a cloudy mood and then headed down to Harborside to meet the gang.

I was the first to arrive, so I sat down by myself. I couldn't even think about coffee this morning. My stomach churned with anxiety, and I knew coffee would make it worse. Soon Dave arrived, then Stefan and his assistant, Ola. I tried to hold in my thoughts about the weather until everyone was there. When Verlin and Bill showed up, I couldn't contain it any longer. The weather had worsened; it was now blowing at least 25 knots of sideways rain. This was about the worst weather I had ever seen in a summer storm on Kodiak. Finally, Josh showed up with Steve Lloyd and their teenage deckhand, Connor. Now there were nine of us, and we took up half the space in the coffee shop. How were we going to get all these people on the boat?

"This weather doesn't look good to me," I said. I started in with one foot then jumped in with the other. "The wind is blowing right in to Monk's Lagoon. It would be dangerous to dive in these conditions and we'll be sitting in the surf all day. I wouldn't choose to go diving today."

There, it was out in the open. Heads nodded in agreement. Josh was a part-time commercial fisherman, and I could see he was ready for anything, but the others were leaning in my direction. We briefly discussed the option of reassessing the weather at noon, but it was so bad that we decided just to put it off until tomorrow. Bill and Verlin, both certified scuba instructors, agreed with my assessment and supported my decision. Steve had brought a magnetometer from Anchorage to help with the search and Josh had wanted to test it out today, so he and Steve decided they would take the boat out for a test run anyway. Dave gave me a quizzical look as if to say, "What's up with that?" then offered to go along with them. Agreeing to reconvene the following morning, we all went out to carry food and some remaining gear down to the *Melmar*, and by the time I walked back to the car I was drenched. But I was relieved that decision was out of the way and that we could use the rest of the day for preparation. As for me, I had to go back to the lab and try to fix a chiller on one of my crab tanks.

I walked back into Harborside thinking that maybe now I'd have coffee. Bill and Verlin were still there. They told me they wouldn't be available tomorrow but could go the next day, Tuesday, if we go. I agreed to keep them informed of our progress. They were probably the most experienced divers in the group, so I welcomed their participation. Now I was finally ready for some coffee, but I had left my wallet in the car. Going across the street to get it meant I would be even more waterlogged when I returned, so I decided just to go home.

Later that afternoon, I drove out to the NOAA warehouse on the Coast Guard base, where I found a crab buoy and line to use for dive markers. Back in town, I visited Kodiak Marine Supply, where I purchased a spool of polypropylene line and two more crab buoys.

When I got back home, I measured out two shots of line fifteen fathoms long, cut and taped the ends, and sealed them with a hot flame. Now we were ready for diving.

MONDAY, JULY 21, 2003

The weather today looked just as dismal. From my house, I could barely see Near Island through the mist, though it was less than half a mile away. Flags in the harbor were still fluttering strongly to the west. This morning's forecast was the same as it was previously, for 15 knots northeast, but it still looked like 20 knots southeast to me. But we couldn't delay this trip any longer, so I put on my long johns. I got so wet just walking down the dock in cotton sweats yesterday that I put on my polar fleece sweatpants today. As a precaution, I dug out my precious stash of seasickness medication, the Coast Guard cocktail. This was a combination of two drugs, ephedrine and phenergan, one of which was now illegal, so I couldn't get it anymore. I would only use them if I got seasick.

As I was getting ready to go, the phone rang. It was Josh, telling me he was having breakfast and would be ready to go in half an hour. After he hung up, I called Dave at his bed and breakfast simply to relay the meeting time, but what he had to tell me was a surprise: Josh and Steve had taken the boat and magnetometer over to Monk's Lagoon yesterday, where Steve dove to the bottom to look around. Dave had suspected they might try to get the jump on us and had gone along for that reason, just in case they found anything.

I immediately had a sinking feeling. That Steve was willing to risk his neck in those conditions was his choice, but I was not comfortable asking a bunch of volunteer divers to do the same. And I was concerned about the fact that Steve and Josh would go there and dive without me. I didn't know exactly where they dove, and it probably wasn't at the right location anyway, but it suggested that they didn't consider this to be a group effort anymore; maybe they never did.

I packed up my gear and went down to Harborside, and the gang assembled shortly thereafter: Josh, Steve, Stefan, Ola, Dave, and me, plus Connor. I grabbed my basket of ropes, buoys, and weights, and we all trooped down to the boat. By nine we were out of the harbor and heading out the channel, right into the wind and waves. If the

The discovery team with the *Melmar*. Clockwise from top right: Josh Lewis, Brad Stevens, Steve Lloyd, Ola, Stacey Beckland, Stephan Quinth; Dave McMahan in center.

swell in the Woody Island channel was any indication, there were going to be some rough seas ahead. I went down into the cabin and wedged myself into a bench seat in the galley. Sure enough, by the time we cleared Spruce Cape, we were banging into 6-foot chop. The fog was so thick I could see neither Spruce Cape nor Woody Island, less than a half mile away. I began to worry. I hadn't been out with Josh before and didn't know how good his seamanship skills were. I began wondering if there were survival suits, where the lifejackets were, and where the EPIRB was. If this were a NMFS charter, I would have required a safety tour and lecture before leaving the dock, but in our haste to get going I hadn't insisted on that.

Josh was driving strictly by his radar, which showed both islands half a mile away on either side of us, but we couldn't see anything through the mist. As far as I could tell, he could have been taking us to Hawaii. But the *Melmar* handled nicely even though we were taking it on the chin constantly, and Josh seemed to know what he was doing. After a while I began to relax. While Josh drove, Dave explained the state's position concerning the *Kad'yak*.

Before the advent of extended jurisdiction, nations controlled only the waters near their shores and as far offshore as they could shoot a cannonball. For centuries that was anywhere from 3 to 12 miles. But in 1977, the United States and many other nations extended their national jurisdiction out to 200 miles, an area now known as the US Extended Economic Zone, or EEZ. That gave them control over a huge area of ocean, and below it were centuries of shipwrecks. The old law of the sea was "finders, keepers." If a ship sank, anyone who found it and recovered the cargo was entitled to keep it, sell it, or return it to the owners for a finder's fee. But in 1988, the Abandoned Shipwrecks Act (ASA) was passed by Congress. The ASA claimed jurisdiction over any shipwreck embedded within state-submerged lands but passed authority for administering the act onto the states with coastlines.

"In order to be covered by the ASA," Dave said, "a shipwreck has to meet three criteria: It had to be abandoned, meaning that there wasn't any remaining insurance claims for it; it had to be within navigable waters of the state—in essence, within 3 miles; and it had to be embedded or partially buried in the bottom, not just sitting on the beach somewhere." The ASA also included wrecks resting on the ocean floor that were listed in or determined eligible to be listed in the National Register of Historic Places (NRHP).

"The Kad'yak," Dave continued, "meets all three criteria. For that reason, it is the property of the State of Alaska and is protected from salvage." He wanted to make sure that all the divers knew that before we dove on it.

After an hour or so of slogging through the waves, I noticed surf breaking to our left—the breakers hitting the reefs on Spruce Island. As we neared Icon Bay, the mist started to lift, and we could make out the islets on either side of the entrance. I hoped Josh could find his way in without hitting a reef. Sure enough, the Melmar slipped right into the bay, and soon we were out of the northeast wind, protected by the cape at East Point.

While Josh drove, I read to him the position that I had estimated for the Kad'yak, and he positioned the Melmar right on it. Three of us pulled out portable GPS units. Both Dave and Josh had new GPS units, and I wanted to compare the readings from their technology with my ten-year-old military unit. Both Josh and Dave's gave slightly different positions, but I trusted mine and it seemed to be working correctly, so I decided to stick with it. When we were as close as possible to my estimate, we dropped a lead weight with a buoy attached to it to mark the spot.

Steve got out the magnetometer and hooked it up. It was a fairly compact unit, consisting of a yellow rocket-shaped towfish about 2 feet long, connected by 100 feet of wire to a yellow metal box with what looked like a voltmeter on it. Steve lowered the fish into the water and paid out the cable. Our plan was to drive around my marker and cover as much ground as possible inside the bay while towing the magnetometer behind us. As we did so, Steve explained how the magnetometer worked.

"The meter should read 0 most of the time," Steve shouted over the noise of the engine. We could hear the magnetometer make audible, high-pitched beeps about every three seconds. "If it detects metal, the beeps will become more frequent and the needle will deflect away from the 0 mark. Any reading from 10 to 40 would be a good hit, and a large metal ship or object might read up to 100."

I had my doubts about the magnetometer. Most of the ones I had seen had to pass within 20 feet of a metal object to detect it. We were towing Steve's at the surface in water 80 feet deep. We knew there would have to be some metal associated with a wooden shipwreck, but how much was anybody's guess. There should be anchors, at least one, maybe more, unless Arkhimandritov had lost them. There were also supposed to be four to six cannons on the ship when it sailed to San Francisco in 1859, but we didn't know if they were present when it sank. And there had to be various pieces of ironwork on the ship to hold it together, but those would probably be small and might even be totally rusted away. Could we really find a wooden shipwreck with this thing? Steve sat on the back deck hunched over the meter to protect it from the misty drizzle. I watched for a while, but the deck was crowded with equipment and I couldn't get close enough to see the instrument clearly. Since there was nothing I could do to help, I retired to the cabin where it was dry.

We must have made about ten circles around the buoy at ever-widening diameters, going out to a quarter mile away on the sides of it, before we noticed the beeps were becoming more frequent. For a few seconds, the needle deflected to 10, then 20, then back to 0.

On the next pass, we got more beeps and more deflection. Josh put a mark on his GPS, and we towed over that spot again. This time the needle went to 40 for about thirty seconds.

"There's definitely some metal down there," Steve said. "But it could be anything. Garbage, an old engine, or just junk."

After making about fifteen circles we decided to stop and have some lunch, so we anchored the boat near the site where the magnetometer hits had occurred. This location was almost exactly midway between my marker buoy and the islet where I thought Arkhimandritov must have stood when he took his bearing to the mast of the *Kad'yak*. If I was correct, the ship would lie between us and the marker buoy, or maybe just beyond it.

The wind had died down now and the water surface was calmer, but there was still a good 5-foot swell rolling into the bay, causing the boat to rock up and down and jerking the anchor chain pretty hard. It didn't bother me, but Connor got seasick and purged into a bucket.

After eating, I decided to dive with Dave and swim toward the marker buoy. Before going in, I stood on top of the wheelhouse and took a bearing with my compass from the boat to the marker—24 degrees magnetic. Dave and I got suited up. I had long johns on already, and over them I put a fleece-lined overall and wool socks, then stepped into the Viking drysuit I had borrowed from Verlin. It was a pretty good fit. I had been using a drysuit owned by my lab for over ten years and it leaked like crazy. By the time I got the Viking suit on, I was sweating, even though the air temperature was only 50°F.

After putting on our drysuits, Dave and I put our buoyancy compensators (BC) and regulators on the tanks and donned our weight belts. My work drysuit was made of neoprene and was slightly buoyant, so I usually carried 40 pounds of lead shot in my belt. TheViking suit was less buoyant, so I took four of the 5-pound weight bags out of the belt. Steve acted as tender, handing us our gear as we needed it and making sure all the straps were buckled and air valves on, and then he helped us over the stern.

The railing of the *Melmar* was a good 4 feet from the water, so to get in I had to balance on my knees on the railing and fall off backwards, landing either on my back or on my feet in the water. My entrance was graceless, landing on my butt and making a large splash, but I came back to the surface upright. The water felt cool and refreshing, though it was actually 50°F, cold enough to cause hypothermia quickly if it weren't for the drysuit. I rinsed my mask out and tested my buoyancy, but I was too light and could not descend. I had to climb up the ladder and grab two more 5-pound shot bags to stuff into my belt. That done, I splashed back in and Dave followed, and we flippered up toward the anchor line.

When we were ready, I let all the air out of my BC and drysuit and sank slowly beneath the swells. I was almost neutrally buoyant, so descending the first few feet took a few seconds, but after that the suit started to compress and I sank faster. As the pressure on my eardrums increased, causing some minor pain, I squeezed my nose and blew, pushing air into my inner ear to equalize the pressure. Then I turned head down and followed the anchor line into the dark water. It was surprisingly clear for July. There was little plankton, and I saw the anchor when it was still a good 10 feet below me. We settled onto a solid rock bottom about 55 feet deep. Even at this depth we could feel the surge from the swells passing over our heads and see the kelp waving back and forth with it. I checked my compass and then swam off at a heading of 24 degrees magnetic.

The rock reef sloped quickly down to a channel bottom at 66 feet. Before entering it, we could see that it was full of cloudy silt moving back and forth with the surge. As I swam into the cloud, visibility declined to almost zero. I could barely see 2 feet ahead in this muck. I'm used to diving in murky silt to chase crabs around in Womens Bay, but this was more particulate, like fine pieces of chopped-up kelp after it's been through a food processor. In order to see the bottom, I had to keep my face within a foot of it, with my hand reaching out to make sure I stayed near the bottom. The bottom here was solid, not sandy or gravelly as I had expected, but paved like a cobblestone street. It was futile to search for a shipwreck in this manner—if I couldn't see any farther than 2 feet, how was I going to find or recognize anything? And how would we ever cover enough area?

For about ten minutes we swam through the soup, and then we came up onto a reef that shallowed up to about 45 feet. The water cleared up slightly, and as we passed over the top we swam through a small patch of kelp occupied by a school of large black rockfish. Male and female kelp greenling were all around the bases of the rocks. They didn't seem overly afraid of us, swimming in close to check us out. On the other side of the reef, we came back down to a sandy bottom and more dense clouds of particulate matter.

The second channel was narrower than the first, and we swam into the outskirts of the kelp forest. I started to get concerned about going too far and getting tangled up in the kelp, but I just kept swimming. Dave was still with me, hanging back just a bit. The bottom shallowed up, the kelp got thicker, the surge stronger, and the muck stirred up by the surge over the reef. When my depth gauge read 38 feet, I decided it was time to ascend before we came up onto the top of a reef with surf breaking over it. I signaled my intention to Dave then put my hand on the biggest kelp stipe I could find and started to follow it upwards. Fifteen feet from the surface, we stopped and waited for three minutes, hanging onto the bull kelp and blowing off the last nitrogen bubbles during a short safety stop. As we waited, I wondered how we were going to get back to the boat. I hadn't asked them to put the inflatable Zodiac boat in the water, but I figured they were probably doing that now anyway so they could come pick us up. At least I hoped they were.

Popping up to the surface, I looked around to find that we had surfaced in the middle of the kelp bed on the opposite side, the north shore of the bay, a good 250 meters from the boat. Looking back at the *Melmar*, I could see that everybody was still on deck and there was no Zodiac coming to pick us up. It would be a long swim back. Dave and I both rolled over on our backs and began kicking our way home. This was a real workout and took much more energy than swimming underwater. If I'd had more air, I would have dropped down 5 feet and swum along underwater, but I was down to less than 500 psi and didn't want to use it unless I had to. Those extra pounds of pressure still might come in handy for buoyancy.

After what seemed like forever, we finally reached the marker buoy. Perhaps Josh took pity on us at that point because he turned on the engine and motored over to pick us up. "Thanks, Josh," I muttered into my regulator, "what took you so damn long?" I was simply annoyed.[3]

Once we boarded the boat, we undressed and related the events of our dive to the others. We were all a bit discouraged. The visibility up above was good, but on the bottom it was miserable. We couldn't see well enough for a decent search, so after some discussion we decided that we would concentrate on the magnetometer site. With that plan, Steve, Josh, and Stefan dove in and swam off to the east while Dave and I rested and drank fluids. The weather gods were being kind to us because the sun had come out, the fog had burned off, and it was quite warm now, though there was still a good 5-foot swell causing the boat to jerk up and down. Ola got seasick and began feeding the fish. Ignoring him, Dave and I went up on the bow, lay down on the deck, and started to snooze in the warm sun.

The second group of divers came back with similar reports of lots of surge and crappy visibility in the channels. The bay seemed to be a network of reefs topping out at 50 to 60 feet, interspersed with deeper channels that went down to 70 to 80 feet. Most of the channels ran east to west and some were like blind canyons, completely surrounded by reefs. The largest channel was paved with cobblestones, but others were filled with course sand or fine gravel.

At about 5 pm, Dave and I decided to make a second dive and swim a square search pattern so we would come back to the boat underwater and not have such a long surface swim at the end of the dive. It was hot now, and by the time I got suited up again I had worked up a good sweat, so dropping into the water was a relief. Usually one dive in these cold waters would tire me out, especially when wearing a drysuit, 40 pounds of weight, and all the attendant equipment. In addition, cold water saps your strength because your body expends energy trying to warm up. However, I didn't feel fatigued now. We dropped down to a reef and swam through narrow passages. The surge was so strong that it was difficult to swim against it, so we timed our movements to work with it, going forward when it pushed us along and stopping when it backed up against us. Down in the sandy channels, the muck was starting to settle, and visibility was a little better, though not by much.

We surfaced to learn that Steve had gone in again. He was using a dual tank setup with twin 100-cubic-foot cylinders so he could stay down for an hour easily. When he finally came up, he called Stefan to bring his camera as he swam back to the boat. Steve came up the ladder and over the rail, finding a seat on a storage box. As Stefan filmed, Steve began to ask me some questions.

"Brad, what do you know about the construction of the *Kad'yak*?" he asked.

"Well," I said, "we know it was built in Germany, it was made of wood, it was supposed to have copper sheathing on the hull to protect it from shipworms, and it probably had ballast stones in it."

"What type of wood was it made of—teak, oak, or something else?" Steve asked.

"Since it was built in Germany it was probably oak, and the ballast was probably granite," I said.

"Did it have any metal on it?" he asked.

"Probably, but we don't know for sure. It would have had anchors, and probably lots of small metal fittings to hold things together. I've heard it also had four cannons on it, but I don't know that for a fact."

"Okay," he said as he looked at me, "well, don't get too excited, but I think I've found a concentration of metal. It just looked like rocks at first, but when I shined my light on it, I noticed it was rust colored, so I scraped on it—and sure enough, it seems to be a mass of metal concretion all rusted together." Then he pulled out of his pocket two small pieces of flat, copper-colored metal. "There are pieces of these all around down there. Lots of them, going down into the sand. The largest are probably dinner-plate sized."

We looked at the scraps of metal. They did look like copper and were still shiny on one side, not green and corroded like I would expect. On the other side they were dark, as if they had been tarred over.

"These could be pieces of the copper sheathing from the hull," I said, studying them. "But it's hard to tell. If there is a lot of it, that would make sense. I can't be sure this is copper though, or what it came from."

Dave looked at them too and seemed pretty certain that it was copper. It was a tantalizing clue. Not enough to get excited about, but encouraging.

We were out of air, so we decided it was time to go home. But first we had to make sure we could find and return to this spot. We made an anchor by putting some unused dive weights into a mesh bag, tying a float and some line to it, and dropped it over the side. Three GPS units were drawn out, and we all took readings of the position. Once that was done, Josh fired up the *Melmar*, and we pulled up the anchor and left Icon Bay, heading back to Kodiak.

The evening was warm, and the waves had died down to a gentle swell, so the ride back was pleasant. We passed around bottles of Corona beer and toasted the day. We may have found something. I felt good. After reaching the harbor, we unloaded all the tanks into my car, and I drove them to Verlin's to get refilled. Then I went back downtown to meet the gang at Henry's Great Alaskan Restaurant, a landmark in Kodiak. Over dinner and drinks, we decided that we would convene again the next morning at Harborside in time for a nine-o'clock departure.

Before going to bed that night, I sat down at my laptop and opened up the navigation program to log the positions that I had recorded on my GPS. The final position where we had dropped the buoy near the source of the copper plates was almost exactly 100 meters southeast of my initial estimate of the *Kad'yak*'s position, and halfway between that and the location of the magnetometer pings. It was damn close. Could it be the *Kad'yak*? If not, then what was it?[4]

CHAPTER 11

ANCHORS
AND
CANNONS

TUESDAY, JULY 22, 2003: THE day dawned clear and calm. I had been waking up hourly all night long. Despite being incredibly tired and with a stomach full of steak, I could hardly sleep. Around six thirty I finally got up, repacked the drysuit I had hung up overnight, grabbed my GPS and backpack, and headed out.

Stacey Becklund from the Baranov Museum was going to join us today, and although I had told her to meet me at Harborside at eight thirty, I was there half an hour early. Dave arrived next, then Stefan, then finally Stacey; but Josh didn't show up. Stefan told us he had gone to Josh's house earlier for coffee but found no one at home. He left to go search for Josh on the other side of the harbor, where we were supposed to pick up the refilled scuba tanks from Verlin. Then as I retrieved my gear from my car, Connor came up the dock saying Josh was on the boat and ready for us. Apparently he and Steve had been there all along, changing the oil. *Okay,* I thought, *that's obviously important, and probably something Josh does regularly.* But it bothered me that he did not communicate what he was doing to me or the rest of the team. Though it should have been a perfectly innocuous event, it bolstered my feeling that Josh and Steve were operating independently of the rest of the expedition.

ONCE WE WERE ALL ON board, Josh started the engine of the *Melmar* and we motored over to the transient float to pick up the tanks. Designated as a place for transient boats to tie up temporarily, locals called it the water dock or the charter dock because it was used by all the charter boats to pick up their passengers and refill their water tanks. Verlin and Bill were there on Verlin's boat, and all our tanks were standing on the dock. We loaded them aboard the *Melmar* and then drove out of the harbor to the fuel dock, where Josh topped off the diesel tanks, before heading out the channel. It was a beautiful morning today, with calm seas and clear skies. Rounding Spruce Cape, we drove into a gentle swell, but it was nothing like the day before. Within an hour we were dropping anchor in Icon Bay next to our marker buoy. Verlin cruised in about ten minutes later and tied up alongside of us. There was still about 3 feet of swell, causing the boats to bang around occasionally, but otherwise it couldn't have been better weather. The forecast said it was supposed to blow 30 knots southwest tonight and tomorrow though, so we had to get our work done here today.

I was eager to make the first dive with Stefan so we could see whatever there was on the bottom together. But before I could get ready, Steve announced that he would be making the first dive. I had a nagging feeling that this wasn't the way things should go, but I didn't have any valid reason to protest. The expedition was a team effort, and credit would go to all. Besides, Steve had seen the site of some wreckage yesterday. He and Stefan jumped in and went down to examine the site. The rest of us waited for half an hour before starting to suit up because we didn't want to get too hot in our drysuits. As we finished, Stefan came up, but Steve was using double tanks and stayed down about twenty minutes longer. When he surfaced, he had a lot to tell us.

"There are some large metal objects down there and some copper rods," he said. "I tied a line from the anchor to several of the objects, and it ends at a large mass that looks like metal all concreted together. When you first look at it, it just looks like a big rock, but if you shine a light on it, you can see the red color of rust. Take lights with you if you can. When you get down there, just follow the line for the self-guided tour of the *Kad'yak* wreck site." Steve had already assumed that we had found the *Kad'yak*. As scientists though, Dave and I were more skeptical and not ready to make that conclusion without additional evidence.

Bill and Verlin followed the *Melmar*'s anchor line while Dave and I descended along the marker buoy line. At the bottom, I followed the line Steve had laid out. The visibility was much improved from yesterday, and I could see 10 to 12 feet ahead even in the bottom of the sandy channel.

The first things I saw were copper rods sticking up out of the sediment. The line led to a large metal tube of some sort, then to more copper rods, then to a large mass of objects all cemented together. I poked at it with my knife, and it felt like rusted metal, but it had rocks protruding from it. On one end there was a shiny mass of what looked like copper, all bent and folded over on itself. For a brief second I wondered about the possibility that it might be coins cemented together, but on closer look, I decided it was just a mass of copper sheeting that had been folded, bent, and pounded together. Next to it sticking out of the mass was what looked to be a piece of kelp, but when I grabbed it with my hand, it felt like leather. A thick leather strap, or hinge, or belt, perhaps. The large mass was about 8 feet long, maybe 3 feet wide, and about 2 feet high, and the top of it was overhung so I could look underneath it. I waited for the sediment that I had kicked up to settle down, then shined my light underneath the edge while poking my face as close as I could into the dark overhang. Briefly, I had visions of a giant octopus reaching out and pulling me in. Instead, I found myself looking at something that appeared to be the end of a wood beam. *Wood?* I thought. That would be a surprise, since most of it should have rotted away.

Groping around near my shoulder, I found my dive slate drifting from a clip on my BC. Dive slates are erasable boards that can be used underwater. With some pencil lead, I drew a picture on the slate, showing the orientation of several metal objects that I had seen, and next to this one, I wrote the word "WOOD" in large letters. Waving my arms frantically, I motioned to Dave, who was hovering nearby, and pointed to my writing then at the beam under the overhang. He nodded in agreement. I pulled out my dive knife strapped to my leg and gently pried off a finger-sized chunk of wood. I would give it to

Dave later for him to determine whether it was really wood and what kind, if possible.

As we swam around the site some more, I came across a long, metal-looking bar that was about 6 feet long, 3 inches in diameter, and curved over at one end to form a J-shape. Then later, I found another just like it nearby. What could these be? Nearby was another large piece of metal that appeared to have a round structure like a wheel or gear attached to it. It was some type of cylinder, about 8 inches across and maybe 3 feet long, but one side had been sheared off, showing the empty hollow inside. I thought it might be a cannon, but it was not thick or heavy enough, and a cannon would not have rusted away so severely. Did they use metal for winches? How would they power it without steam?

It was time to go by then, so Dave and I surfaced along the line and swam back to the boat. At 15 feet we made our three-minute safety stop, and while hanging there in midwater, I reflected. There sure was a lot of metal down there. What were all those copper rods? The large concreted mass? The curved iron bars? The thing with the wheel? This wasn't just junk somebody threw overboard; there was a pattern to the distribution of stuff, although I couldn't tell exactly what it was. There were enough pieces spread out over an area just about the size of a ship. It had to be a shipwreck site; that was the most sensible explanation. My earlier feelings of encouragement slowly evolved into a feeling of satisfaction. We were definitely onto something. But was it the *Kad'yak*? I still could not be sure.

Bill and Verlin were already back sitting on the stern of his boat. After I climbed aboard and removed most of my gear, Bill called me over.

"Brad, do you want to hear the bad news or the good news first?" he asked, although his smile suggested that whatever the bad news was, the good news was going to be way better. To describe Bill as laconic would be hyperbole. He was about as low-key as anybody I had ever known. I had never seen him ruffled or excited, but now he was eager to tell me something.

"We had a long swim back," he said. "We came up way over there near the kelp bed. Now you want the good news? We found a cannon."

My jaw dropped. "A cannon? Are you sure?"

"I'm sure it was a cannon. It was about 3 feet long, and it had a rounded butt with a nub on it. The end of it had a hole with a bore about the size of my hand," he said. Verlin nodded in agreement, as he had seen it too.

I couldn't believe it. A cannon! Most wreck divers would give their eyeteeth to find something like that. It would definitely date the wreck and might even have writing that identified its source. But it was still at the seabottom, not in our hands.

"Can you take us back to it?" I asked. They thought they could. They had descended the anchor line and swam 240 degrees for about 100 feet. At the least we needed to know where it was relative to the other wreckage, but Bill and Verlin had not seen any of the other artifacts on their dive, so we couldn't know exactly how the cannon fit into the scene without relocating it. And I was worried that if we waited, winter storms might bury it in the sand where we couldn't find it again. We needed to find it again, and maybe even recover it.

We decided immediately that we would send out a search party, but first we asked Dave about possibly recovering it. He was skeptical, but he got out his cell phone to call his office in Anchorage. It took several tries to get a clear signal from our location, but finally he got through. We waited anxiously while he discussed the situation with his supervisor. Finally, he ended the call and told us the news. We had the go-ahead to attempt recovery of the cannon. If we could find it.

Steve, Josh, and Stefan made the next dive since Dave and I needed time out of the water to blow off nitrogen. We munched on sandwiches that Stacey had made. *This is it*, I thought. It had to be the *Kad'yak*. With a cannon. And where there was one cannon, there might be more.

About an hour later, we saw a yellow lift bag pop up to the surface about 100 yards away. Steve had found an anchor. He had tied the lift bag to it and inflated it to act as a buoy marking the spot. An anchor! First a cannon, then an anchor. We couldn't believe our luck. I never expected to find so many significant artifacts in such a short time. Perhaps after surveying the site completely and sifting lots of sand, we might have found such things, but I had never in my wildest dreams believed it would be this easy.

But they did not find the cannon again. We needed to have as many eyes in the water as possible, so I asked Bill, Verlin, and Dave to accompany me on another sweep. If we spread out to the extent of visibility, we should be able to cover a fairly wide swath of bottom. During our two-hour surface interval, the wind had come up and was now blowing from the southeast, right into the bay, kicking up a 3- to 4-foot chop. The boats were banging together, and to make things worse, Josh discovered we had dragged anchor and were drifting toward the reef behind us. While Connor and Steve raised our anchor, Josh fired up the *Melmar*'s engine and began to move us slowly forward, back to where we had been, towing Verlin's boat alongside. In the chop, the boats began knocking together furiously. Lines tying the boats together became taut and started to snap. Suddenly one of Verlin's cleats broke loose. He and Josh ran out to untie the boats, but the line was too tight and they couldn't loosen it. More rocking and banging occurred until someone finally found a knife and cut through the line. Verlin started up his engine and moved away from us some distance.

We had to find the position where we had previously anchored. I had not recorded it on my GPS because I was using the marker buoy for a position. Josh thought we had been in 72 feet of water. For the next few minutes, he drove the boat from the flying bridge while I sat down in the wheelhouse calling depths out to him through the window. Eighty feet over the sandy channel, then up to 45 over a reef, then edging back down to 60, 65, then 70. We dropped the hook and shut down the engine. Looking back at the marker buoy, I could see that it was farther away from the boat than it had been earlier, so we couldn't be at the same location. But Josh thought we were, so we suited up for diving and plunged in.

Before descending, all four of us converged at the anchor line. I asked Verlin to take the lead because he was fairly certain of his compass bearings. Bill went on his right, and Dave and I were on his left. Down the line we went to find our anchor sitting on a reef in about 65 feet of water. We dropped to the channel bottom, spread out in a search line, and oriented ourselves to the west. Then we swam off at a bearing of 240 degrees searching the bottom. We were in a wide channel with a sandy bottom and lots of scattered rocks. The visibility had improved greatly, and I could see now at least 25 feet away. We swam slowly, searching for any object that might be a cannon. After about 50 yards, I saw a furrow in the sandy bottom leading away from the reef on the left and out into the open channel. It was a drag scar from Josh's anchor that had dragged through the sand. That meant that we had previously anchored near here, about 100 feet closer to the wreckage than we now were. *Perhaps we should've searched on the other side of this reef rather than in this channel*, I thought. At the end of the drag scar, the channel narrowed and split into two smaller channels on both sides of a large rock. We swam to the left of the rock and found ourselves back at the spiderweb of line on the main wreck site, next to the marker buoy. There we stopped.

Verlin motioned for us to turn around and search the channel again. He apparently thought we had gone too far. We all followed suit, and as we swam back, I saw to the left side of the channel near the base of the reef something that looked like a log. It turned out to be a large metal spar or beam of some kind, about 20 feet long and 6 inches in diameter. Next to it was a short piece of copper rod; I picked it up and put it in my goodie bag, but it was extremely heavy.[5]

We didn't see any other wreckage in this channel and soon found ourselves back at the boat anchor. Once more, Verlin motioned to turn around and go back through. We started on a third search of the channel but didn't get far before Dave signaled to me that he was low on air, so the two of us surfaced while Bill and Verlin continued the search. They surfaced not long after us. Nobody had seen the cannon. It wasn't going to be found. What should we do next? Bill and Verlin were out of tanks, so they decided they would head back to town.

Josh and Ola were the only ones who had not made two dives, and Steve was ready for a third. We didn't want to waste any more time searching for the cannon, so we decided to go back to the anchor that Steve had marked and film it. Given our options, that seemed like the most reasonable thing to do. From our present position to the float bag marking the spot was a good 100 yards, so Josh moved the *Melmar* near it and re-anchored. Placing me in charge of the boat, he told me to watch the reef behind us; if we started to drag anchor again, I was to pull it up and move the boat.

After they entered the water, I went up on the flying bridge and placed my GPS on the bench. Not only did I want a good fix on our position, but I wanted to make sure we did not drift. Sitting up there, I checked the position every few minutes, watching the digits. The wind was blowing from the southeast. As long as our longitude didn't change, we were fine. The shallow reef was right behind us, to the west, so I watched the GPS numbers like a hawk. They didn't budge. The anchor was holding.

In fact, the wind died down and it became calm again. Except for rocking in the swell, it was downright balmy. I joined Stefan, Connor, and Stacey up on the bow, where they were watching a live video feed on Stefan's camera that was coming up from the Splashcam carried by Steve. On the 3-inch screen, we could see him pointing it at the anchor, swimming around it to give us a better look. Looking up, I could see his bubbles coming up near the lift bag. I took out my hiking compass and took a bearing toward the marker—168 degrees, at a distance of about 100 feet. Another bearing to the original marker buoy was 103 degrees at a distance of about 200 feet.

Now we saw on the screen a large, barrel-shaped object. It was perhaps 3 feet across with thick walls and a 2-foot hollow inside. Steve held the camera in the hollow, and we could see worms and other creatures growing on the walls. What was it? The picture turned upside down briefly as Steve laid the camera on the seafloor. Soon he picked it up again, and we could see an anchor. But wait, this one was smaller than the first one we had seen. It had to be a different anchor! We watched as the camera moved over rocks and sand before focusing on a large, cylindrical object. What was that? Suddenly we realized we were looking at the cannon. They had found it again! Up on the deck, we were hooting and hollering, excited and happy that it had been relocated. Was it the same cannon or a different one? It looked larger than the one Bill had described, but it was difficult to tell without some frame of reference. Even if it wasn't, I couldn't believe our luck. Two anchors and a cannon on the same day. What were the odds of that?

Getting out my waterproof notebook, I drew a sketch of the boat, the two marker buoys, their compass bearings and distances from the boat, and the location of Steve's bubbles. He was at the cannon, and his bubbles would tell us how far away it was.

Surprisingly, it was right about in the middle of the two buoys. That meant we had not gone far enough during our last search. If we had swum 50 feet farther, we might have seen it. Dave and I might even have passed by it this morning but not seen it in the low visibility conditions that were present then. At any rate, I now had a map that would bring us back to the site, no matter what.

Soon Josh and Ola came up, but there was still no Steve. We could see his bubbles move off toward the first marker buoy, but they did not go any further. After about ten minutes, we concluded he must be doing decompression stops. This was his third two-tank dive. He had been down at 80 feet over forty minutes, so he had to be in decompression. Indeed, that was what he was doing, as we learned when he finally surfaced and swam back to the boat. After climbing aboard, Steve reached into his collecting bag and pulled out some metal objects. One was a flat metal strap or bracket, probably bronze, about 3 inches wide, half an inch thick, and 10 to 12 inches long, with two foot-long, inch-thick bronze pins inserted through it. The others were shorter versions of bronze pins. Both Steve and Ola verified that the cannon was quite close to all the other wreckage we had seen, and we were all surprised and chagrined that we hadn't seen it earlier.

It was now eight thirty in the evening. It would be impossible to recover the cannon today. There was a storm brewing for tomorrow, so this had to be the end of our efforts. We pulled up our marker buoys so nobody else could find the location of the wreckage, and then we headed back to town.

We all felt many emotions—excitement, elation, happiness, satisfaction. On the ride back, we were all smiles. Stacey couldn't wait to tell the museum board the good news. I didn't know who I would tell. There was little doubt in my mind that we had discovered the wreck of the *Kad'yak*. Two nineteenth-century anchors, a cannon, and numerous other artifacts littered the site. It was within 100 meters of my first estimate of its position, easily within the margin of error caused by the difficulty in interpreting Arkhimandritov's bearings and the locations from which they were taken. I thought he had taken the bearing to the mast from the third islet, which I could see now was no more than a large rock jutting up barely above the high tide level. If he had taken the bearing from the second islet

Steve Lloyd shows off the first artifacts discovered.

a bit further to the west, his bearing, as I calculated it, would have run right through the location of the wreckage that we found. It had to be the *Kad'yak*.

Earlier in the trip, I had promised that if we found the wreck, I would ring the bell on the *Melmar* and buy everybody a drink. So I did just that. In Kodiak and perhaps other towns where people make their living from the sea, every bar has a ship's bell in it. Patrons who were in good spirits, had some extra money to burn, or just came in with a fortune in fish or crab would ring the bell to announce that they were buying a drink for everybody in the bar. Tourists sometimes would ring it out of curiosity, then find themselves with a hefty bar tab when all the locals ran up to get their drinks. Today, it was my turn to ring the bell. But only on the boat, not in the bar.

We arrived back at the water dock around 10 pm and offloaded all the scuba gear and hosed some of it down. This being summer in Alaska, it was still light out. After Josh put the *Melmar* back into its slip, we all met at Henry's and I bought them the promised drinks. As we sipped, we discussed our options. Should we try to recover the cannon at a later date? I was worried about looters pilfering the site. Sure, it was remote, and you needed to be an experienced diver to get down there, but there were some divers around town whom I did not trust. Would winter storms bury it where we couldn't find it again? Should we tell anybody? Before separating, I swore everybody to secrecy on the location of the wreck. But I didn't mind them telling their families we had found it. Maybe that wasn't the best choice, but I knew the cat was not going to stay in the bag very long in this town. The "bush telegraph" was often the fastest form of communication, if not the most accurate. More to the point, a lot of people had contributed to our search, including Mike Yarborough, Lydia Black, and others. I didn't want news of our discovery to become public in a way that didn't acknowledge the contributions made by everyone who was involved. So perhaps it would be best to make a public announcement soon so that we could emphasize that the State of Alaska now had jurisdiction over the site and any disturbance would be illegal. Many people knew the story of the *Kad'yak* and that it probably lay somewhere in Monk's Lagoon. Yes, it would take a lot of effort to find it without the historical documentation, but somebody might get lucky, just like we did.

Dave and I drove back to the lab, where we rinsed off all our gear and hung it up to dry, after which I took Dave back to his B&B. As he got out of the car, Dave paused.

"I'm wondering about something," he said. "When Steve came up from his last dive, he had two sets of bronze pins with him, which I asked him to turn over to me when we got back to town. But he only gave me one. I'm hoping he just forgot, and the other one is still in his gear bag." He shrugged. "But I don't want to make too much out of it, so I'll just ask him tomorrow. Have a good night."

I was too tired to absorb it. Some artifacts were missing, and Dave was concerned. Should I be? Steve had made a big contribution to this effort. He not only brought the magnetometer with him, but he was the first to find metal and to locate the anchor and cannon definitively. Did he think that entitled him to keep a piece of the wreck? *That couldn't be possible*, I thought; I had to believe it was just an oversight.

I drove home but was too excited to go to bed. I was high on life and on adrenaline. Even though it was past midnight, it was still warm and still dusk. I opened a bottle of Spanish port, sat out on my deck overlooking the harbor, and drank two glasses. What an incredible day it had been. And yet, something didn't smell right.

CHAPTER 12

THE
SHIP
HITS THE
FAN

WEDNESDAY, JULY 23, 2003: THIS morning, I spent several hours recording dive logs on my computer and entering the positions of our dives and some artifacts into my navigation program. As soon as I could, I sent an email to Tim Runyan and his student Evgenia (Jenya) Anichenko at ECU. I wanted to tell them the great news. Along with the email, I sent a photo of the metal bracket and pins we had recovered from the wreck. It had been sitting in a bucket of freshwater overnight. Stacey had asked me to bring it to the museum today, so that was what I planned to do.

I had been away from work for several days, so it took until noon to get the shipwreck out of my head before I could think about my real job again.

THURSDAY, JULY 24, 2003

I still had not heard back from Tim or Jenya, so I placed a call to Tim's office. His secretary told me he was in Jakarta but gave me Jenya's phone number. I called her at home and delivered the good news, and she was ecstatic, saying she had already received an email from Dave McMahan.

Her husband and fellow graduate student, Jason Rogers, was also on the phone, and he expressed his concern that we had collected the bracket without recording its position first. He was right; in our excitement at finding the ship, we had allowed Steve to recover objects without first documenting them. He told me to keep it in saltwater until it could be slowly desalinated, and asked if we had told anybody about the wreck. I had to enlighten him about the bush telegraph and how information often travels around small towns faster than one imagined. After finishing our phone conversation, I sent them an email with full details of the discovery. Then I went downstairs to put the artifacts back into a saltwater tank in the laboratory. But Jason's comment still stung. We had done something amateurish, and I felt like a dunce. I would not let that happen again.

FRIDAY, JULY 25, 2003

Did I really think I could go back to work and get anything done? There was a manuscript on my desk that I had been trying to get off to the publisher for a month. All it needed was one final proofreading and a cover letter. When I came into my office this morning I had every intention of finishing it up and sending it out in the mail, but it was not to be. Lurking in my email was a bombshell about to go off.

Usually I spent the first half hour of my day reading email, then went down to the seawater lab to check the crabs, check the water temperature, and talk to my lab associates. But upon opening my mail today, I saw a message from Steve Lloyd. The heading was ominous: "Historic Shipwreck found." When I opened the letter, my heart sank. It was a news announcement dated today with two full pages describing the details of the wreck discovery and Steve's name as the author. Furthermore, it stated that the wreck had been found by Shoreline Adventures, LLC. What the heck was that? As I read it, my pulse quickened, and I sank into my chair. Steve's announcement had given out virtually all the details except the GPS position. He wrote that the *Kad'yak* had sunk in Monk's Lagoon (of course, that was in the historical record, but there was no need to confirm it) and that we had seen anchors and cannons. Just the sort of thing that would be enticing to pillagers. I just couldn't understand why he did that. I thought we had agreed not to publicize the discovery until later. I hoped that this was just a draft and that he hadn't actually sent this out to anybody yet.

I responded immediately to his email, thanking him for his assistance and asking if he had given that announcement to anyone yet. Then I tried calling him at his business in Anchorage, Title Wave Books, but he wasn't in. So I tried Josh next, who had gone up to Anchorage and told me that the story had gone out to the *Anchorage Daily News* and the *Juneau News*. He seemed surprised that I would object to releasing the information. He also explained that Shoreline Adventures was a company he and Steve had formed to find and salvage shipwrecks. I was flabbergasted. I didn't even know they had a company, much less that they were in the business of salvaging shipwrecks. Is that why they were in this, so they could file a salvage claim on it? What a naïve dolt I was—I had been totally blindsided.

Immediately, I called Stacey at the Baranov Museum and told her what I had learned. Of course I was upset that Steve and Josh were claiming all the credit for the discovery, but my major concern was that people in Kodiak should not have to learn about it from the Anchorage newspaper. That would not be right, and it could really upset some people. Residents of Ouzinkie, in whose collective memory these stories have lived for over a century, should learn about this event from local sources. The Baranov Museum directors, who had agreed to help fund our trip, should learn about it from us directly. It had been my hope that we could organize a single press conference at a later date at which all the principals could be present and recognized for their contributions and we could put out a single story and some photographs and video. Having this information come from several different sources at different times would create nothing but confusion. Stacey and I knew we had to react right away, but how? We decided the best thing to do was to organize a press conference for that afternoon at the Baranov Museum.

After hanging up, I was visited by a graduate student who needed help with her statistics. I gamely tried to help her, but it was no use. Ten minutes later Stacey popped into my office with the news that Adam Lesh, editor of the *Kodiak Daily Mirror*, could meet with me at the museum and put a story in the *KDM* that afternoon. That was the end of my attempt to have a normal workday. I apologized to the student and signed out for the rest of the day. At the museum

I met with Adam, who took notes for a brief and vague story. He was quite considerate of our situation and even held the press up for an hour so he could put the story together. An hour later a small crowd assembled at the Baranov Museum for the announcement, including reporters from KMXT Public Radio, the *Kodiak Daily Mirror*, independent journalist Mike Rostad, Lydia Black—whom I specifically invited—and a few of the museum's directors. Stefan, Ola, and Bill Donaldson also showed.

A few minutes later, Adam returned with Friday's copy of the *KDM* bearing the headline "BREAKING NEWS: Shipwreck Found" (see Appendix B.1). I had never seen the phrase "breaking news" used before by the *KDM*. To the assembled crowd, I described how I had received the *Kad'yak* information from Mike Yarborough, the difficulty of interpreting the bearings, my epiphany upon seeing Lydia's chart, and then the two days of diving at Spruce Island. I never mentioned Monk's Lagoon, the depth, or the presence of anchors and cannons. I did, however, display to the onlookers the metal bracket and pins that we had found. Then Stefan showed a few minutes of video he had prepared; even though it showed the anchor and cannon, I didn't draw any attention to it. Those present could see for themselves, but we wouldn't confirm it in print or on the air.

Afterwards, Lydia Black invited me to come to her house for a quick visit. Together, we sent an email to Kathy Arndt at the University of Alaska, who had translated Arkhimandritov's log that I had used to find the *Kad'yak*. We wanted to know if she could locate accounts of the wrecking or reports of any attempts to salvage the ship. Arkhimandritov must have filed some report with the RAC. Where was it? Lydia also made a phone call to Dr. Richard Pierce, who had studied and written about the *Kad'yak* and was now retired and living in Ontario. I spoke with him briefly about the discovery, and while doing so, his wife got on the phone and said what a nice birthday present this was, because he would be eighty-five tomorrow. I asked Dr. Pierce what he knew about a chart showing the *Kad'yak's* position. An article in the *Kodiak Daily Mirror*, published back in 1986, showed parts of this old Russian chart, with the words "Barka Kadyaka" written on it, accompanied by an interview with Dr. Pierce about the chart. But he didn't recall the story or the chart and couldn't help us locate it. I promised to send him a copy of the news article, hoping that it would help jog his memory.

Before leaving Lydia's apartment, I told her about the dream that I had on Spruce Island, about the old woman in strange clothing speaking in an unknown language who had suddenly appeared out of the forest and then just as quickly disappeared.

"Oh, yes," she said, with a slight twinkle in her eye, "that was Mary, Father Herman's closest associate. She wanted to help you find the *Kad'yak*."

It made perfect sense: My dream had amalgamated pieces of information that were drifting around loosely in my brain and connected them into a semicoherent picture. But wait—could the woman in my dream have been Lydia?

I left Lydia's place in a daze, totally consumed by the events of the day and Lydia's interpretation of my dream. On autopilot, I meandered down to Henry's to meet Stacey. We both needed a drink, and margaritas fit the bill. It had been a crazy day. I was torn by different emotions. On one hand, I felt as if I had violated the security of the site by announcing our discovery, but on the other hand, even more information had been given out to the *Anchorage Daily News*, and we were now trying to corral the story and refocus it on Kodiak, where the events occurred. It was a Hobson's choice. Neither action nor inaction was the best course, but action was clearly better.

At five o'clock, Stacey and I went out to my car to listen to the KMXT evening news. They did a good job with the story, narrating most of the important information, and including a few bits of live conversation in which I described the artifacts recovered and talked about the ship construction. While we listened, a feeling of resignation fell over me. This wasn't the way things should have happened; we should have held a joint news conference. But Josh and Steve had already released their own version of the story, claiming credit for the discovery. The only thing left to do was try to counter their narrative. We had done the best job we could, and would have to live with it.

SUNDAY, JULY 27, 2003

This morning, Bill, Verlin, and I were going back to the *Kad'yak* site to do some underwater photography, with Dave's blessing, of course. Stefan had taken lots of video footage, but we didn't have any decent photographs of the site or its artifacts. At seven, I was out the door, and after stopping at Walmart for camera batteries, I drove to Verlin's shop, where we loaded up our tanks and other gear.

The last two days had been fairly calm, so today should've been a perfect day for diving. But the fog was so thick this morning, we couldn't see very far ahead. As we headed out the Near Island channel at a snail's pace, Bill and I stood forward lookout on the bow. It took us an hour and a half to reach Icon Bay, and even when we pulled in, we couldn't see the shoreline. I was totally disoriented. Standing on the rear deck with a GPS in one hand and a compass in the other, I directed Verlin.

"West," I said. "Now south. No, stop. Go 10 meters more to the east. Stop. Now more south. That's good, drop the anchor." The splash of the anchor was muted by the fog, giving it a creepy, otherworldly aspect. Kelp bladders floated on the still water at the edge of our vision, like tentacles reaching up from the reefs to grapple the unsuspecting boat. They suggested the closeness of sharp rocky reefs that lay unseen in the fog.

My major concern was being able to find the location of the wreck repeatedly. I trusted my GPS to get us back to the spot, but I wasn't sure that it would read the same as any other one. By placing my unit on the deck beneath the antenna for Verlin's GPS, I could see that we were both getting the exact same readings. That was a good sign. We dropped anchor near the location where we had first encountered parts of the wreckage, and as we let out the anchor line, the boat came to rest exactly above parts of the wreck. Islands and shoreline soon began to appear out of the fog, providing us with dim views of land. As the fog lifted, I could recognize landmarks again and saw that we were indeed in the right location. Technology is great when it works, but if it doesn't, you don't want to trust your life to it.

We all suited up. Verlin stepped into the water first and I followed. Bill handed us both our cameras then plunged in. I was carrying my Nikonos 35mm underwater camera with two flash units, one wired to the camera and the other acting as a slave, and a 15mm "fisheye" lens. Underwater, its field of view was reduced to something more like a 20mm wide-angle lens in air. The best thing about it though was that the short focal length has a very wide depth of field. I could focus it at 1.5 meters, and anything between 1 foot and infinity (about 30 feet underwater) would be in focus. Of course, the light from the flashes would not penetrate more than about 10 feet underwater. Their primary use was to illuminate colors, particularly red, that are absorbed by water and disappear below about 30 feet.

We dove down along Verlin's anchor line, landing on the rocky reef at about 65 feet. Bill pointed behind me, and I turned around to find a 6-foot halibut lying on the reef. My first thought was *Wow!* It's rare to see a fish that size while scuba diving. My next thought was *Where's my speargun?* which quickly faded away. Every dive can reveal something new and exciting, and the important thing was to observe, absorb, and appreciate it. Even if it wasn't what we came for. The halibut stayed only long enough for me to take its picture before exploding up and away, leaving a cloud of sediment in its place. Turning around, I followed the reef down to the sandy bottom and looked around to orient myself. The visibility was almost 30 feet, much better than we had previously seen. I swam around to find the string tied around a rock that marked the beginning of the wreckage.

My goal on this dive was to rediscover all of the major artifacts that had been seen by each of the divers. None of us had seen them all and we wanted to get a visual orientation to the wreck site. We followed the string to the large mass of rocks from which I had picked out a piece of wood embedded with mashed copper. I picked up a rock nearby and noticed that it was light gray, like granite—not blackish-gray, like the slate making up most of the Kodiak rocks. Realizing that this large mass was probably a pile of ballast stones, I carefully deposited it back where I had found it.

Nearby, we found a 4-foot-long metal object shaped like a piece of pipe that was sheared off longitudinally, exposing the inside. At one end were two round objects, about a foot in diameter, which looked like pulleys or wheels but were not connected to each other. Nearby was a cannon, about 3 feet in length, lying on top of some rocks. A small part of the muzzle end was chipped away, and the bore was

Kad'yak artifacts found in July 2003. Top: Bill Donaldson hovers over a cannon. Bottom: The main bower anchor. Note sea anemone on left fluke.

full of gravel and sand. Beyond that was a curved piece of iron, about 6 feet long and 3 inches in diameter. It looked like it might have been a davit for a lifeboat, and I had seen similar objects in photographs of contemporary ships. The fact that I had seen this on one of my previous dives made me realize that I must have swum right over or damn near the cannon on that same dive. How had I missed it?

We swam onward in the same westerly direction for another five minutes without seeing any more objects, coming to a stop at a location where the sandy bottom gave way to rocks and started to shallow up towards the base of a reef. I tried to write the word "anchor" on my slate, but the pencil lead broke and it became illegible, so I drew a picture of it instead. Bill apparently understood and mimicked the shape of an anchor with his hands to Verlin. Using my compass, I veered around until I was facing out into the center of the channel, then pointed in that direction as I swam off toward the south. Within two minutes an anchor appeared in front of me. It was much larger than I expected. Underwater, it looked about 8 feet long. It had large flukes about a foot across, one of which a large sea anemone had claimed for its home. This had to be the *Kad'yak's* main anchor, or bower anchor. I realized then how heavy it was and how difficult a job it would be to retrieve it.

We had used up about half of our air, so it was time to begin taking pictures. I took several of the anchor, some with Bill in them to give a sense of scale. Turning back to the east, we soon saw the round barrel-shaped object we had seen in the video. Now that we were up close, we could see that the hollow inside was lined with what looked like remnants of wood. Next to it were three large links of chain, each about 8 inches long and 6 inches across. While I took several photographs of these, it occurred to me that this barrel-shaped object might be part of the anchor windlass, the manual winch used to haul up the anchors and to take in heavy mooring lines. Its shape, the presence of chain links, and proximity to the anchor all seemed to support that conclusion. Not far away, we found the second, smaller anchor, which was probably a kedge anchor. It was about 5 feet long and not nearly as massive as the bower anchor. Kedge anchors were simple devices made of welded round iron rods and used to adjust the ship's position or to pull against wind or current. From there we swam into a void where no wreckage could be seen, but after another minute we found ourselves back at the main wreck site. I took photographs there of the cannon, davit, metal objects, and ballast pile.

As an experienced diver, I would usually monitor my depth, time, remaining air pressure, and buoyancy at regular intervals when I'm underwater. It's a lot to think about but has become second nature after years of diving. However, it can be easy to get distracted underwater, especially when photographing sunken artifacts or just admiring the amazing scenery. Such "task overload" can lead to trouble.

Suddenly I remembered to check my decompression computer. I had been down in the water for thirty minutes and had two minutes of bottom time left but only 250 psi of pressure left. That was barely enough for a safety stop. I had better head for the surface right now! I swam over toward the reef, starting my ascent at the same time. I didn't like the idea of surfacing without a guide line because I didn't know where I would come up, and it was difficult to gauge my rate of ascent without a visible reference, but I didn't have any other choice. About halfway up, I spied the anchor line and headed toward that. Fumbling with my camera, drysuit valve, and depth gauge, I lost control of my buoyancy. Air in my suit expanded like a balloon and I started rocketing upward. I couldn't dump air from the suit fast enough. Looking up, I could see the water surface above me and knew I had to stop my ascent. If I surfaced too quickly, air in my lungs would expand too far and cause blood vessels to break, giving me an embolism. That could be deadly. I turned completely upside down and started kicking downward, struggling against the direction of my ascent, until I came to a stop about 10 feet below the surface. I continued down another 5 feet and grabbed onto the anchor line. My pressure gauge was so low it was unreadable, but I was still breathing. I stayed as long as I could, about two more minutes, before I ascended and popped up in front of the boat.

After climbing in and taking off my gear, I realized I was not wearing my ankle weights. That had made me about 3 pounds too light, allowing air to pool in my ankles and contributing to my inability to control my ascent. Fortunately, I had stopped in time.

Kad'yak artifacts found in July 2003. Top: The ballast pile, a group of granite rocks cemented together. Coppery parts are bits of metal. Bottom: Ballast pile in 2004, showing exposed ribs and hull planks which were buried under gravel and broken shell in 2003. *(Bottom: Tane Casserley/NOAA.)*

I cursed myself for being careless and swore to watch my gauges more closely on the next dive.

Other than that, it had been a successful dive. I had shot up an entire roll of film with no camera problems. Bill and Verlin weren't so lucky. Bill had accidentally pushed the side of his mask inward as he was adjusting it, causing it to flood. And Verlin's camera refused to function for some unknown reason. He couldn't explain the failure, but upon taking the camera out of its housing, shooting a few frames, and putting it back in, it started working again. Even experienced divers have bad days.

We sat, ate sandwiches, and talked for a couple hours while we waited to blow off nitrogen from the dive. The fog had disappeared completely, revealing a beautiful sunny day, but the wind was starting to come up from the south, kicking up some chop. The forecast was for 10 to 15 knots variable, which could mean anything from no wind to 20 knots southeast. It was beginning to look like the latter condition was more likely.

My plan for the second dive was to take up slack in the string laid out by Steve and tie together as many of the artifacts as I could, making a guide line so that I could find them all again, or could find my way back to the boat directly. Then I planned to record the bearings and estimate the distances between all the artifacts. Verlin had more nylon twine on a spool in his boat, so I stuffed that in a mesh bag hanging from my weight belt. I wanted to extend the existing line all the way out to the cannon and anchors, if possible. We plunged back into the water for our second dive and followed the anchor line back down once more. This time, Bill and Verlin swam off to take pictures. They knew I would be on the guide line for the whole dive, and I was comfortable diving solo, so they took off together, leaving me to do my work.

This was an unwritten secret of scuba divers. One of the most important rules learned by novice divers was: Never dive alone. If you get into trouble or have a medical emergency, you could die. But experienced divers sometimes preferred to dive solo, especially if they were trying to do photography or some other work that would be difficult with other divers around kicking up the sediment. My employer, NOAA, had strict rules about staying in buddy teams, which we obeyed, but this was not an NOAA operation. Today we were using the SOB system, or Same Ocean Buddy system—if we're diving in the same ocean, we're dive buddies, right? This was, of course, somewhat tongue-in-cheek, as we were not that far apart. Still, I would *never* recommend this to novice divers. It was my 441st lifetime dive, and I remembered my ankle weights this time.

The increasing wind and seas stirred up the sediment all the way down to the bottom, and the visibility decreased to about 20 feet. I located the beginning of the guide line where it was tied to a small rock, but since that wasn't part of the wreckage, I removed it and retied it around a metal rod sticking up out of the sand. Even if you know how to tie knots well, tying them underwater isn't the easiest thing to do, especially with big Gumby gloves and a bit of nitrogen narcosis, so it took me a few minutes to figure it out. After taking up the slack in the line, I swam to the next object where I untied and retied the line again. I repeated this until I reached the end of the first line at the large metal object. There, I tied the second line into the first, swam to the cannon, and wrapped the line around it several times. I continued onward and repeated that task at the davit, finally running out of line at the kedge anchor. With such good visibility today, I could see the windlass in the distance, and from it the bower anchor was just a short swim.

Starting from the anchor, I took compass bearings and estimated the distance between subsequent artifacts as I swam back to my starting point. The stock of the anchor pointed almost directly north toward the barrel-shaped object, to which I estimated the distance at 30 feet. From there, I took another bearing as I swam to the small anchor. I could not see any of the other artifacts, so I had to swim along my guide line before the cannon came into view. At each artifact, I took a bearing along the guide line until I reached the starting point. After writing the last bearing on my slate, I checked my dive computer and saw I had three minutes left and had sucked my tank down to 250 psi again. This time I knew that was enough for the ascent, so I kicked leisurely toward the surface, stopping for a two-minute safety stop. I surfaced into a 3-foot chop that threatened to bang me into the hull of Verlin's boat, so climbing back up the ladder by myself was an adventure. Verlin and Bill came up right behind me. We briefly debated having a third dive, but we would have had to wait two hours or more before jumping back in to have any bottom time. We were now bouncing around in the sea and the wind was getting stronger, so we decided against it. Bill pulled up the anchor and we headed back to Kodiak.

I was anxious to see the outcome of my photography but also conscious of the fact that it had taken a tremendous amount of effort to obtain the photos. It would be difficult, if not impossible, to repeat. In 1992 I had lost some valuable photos of crab aggregations taken from the *Delta* submersible (at a cost of over $100,000) because the film was lost by the local drugstore that was responsible for shipping it to an off-island processor. Today's film was Fujicolor print film, so it could be processed in one hour at Walmart. I had taken plenty of film there in recent years, and they usually did a good job with it, occasionally even printing them in focus. The *Kad'yak* wasn't going anywhere, so there was always the possibility of returning. After weighing the risks and finding them acceptable, I drove to Walmart and dropped the film off for one-hour processing, even though it was after 7 pm and I wouldn't get it back until the next day.

Almost as soon as I got home, the phone rang. It was a reporter for the *Anchorage Daily News* calling to inquire about Steve's story. Of course they wanted photos and details. They had been given some photos by Stefan and asked me about them. One thing I wanted to be sure of was that they didn't show any photos of the anchor or cannon because I thought it would be too enticing to looters. In that request, she accommodated me, agreeing only to print the photo of the barrel-shaped object, what I had suspected was the windlass.[6]

MONDAY, JULY 28, 2003

I went to work this morning once again thinking that I was actually going to accomplish something. Upon arriving, I learned that the *Kad'yak* story was the front-page headline in the *Anchorage Daily News*, right above news of the Iraq War (it was becoming that stale). Email was flying in faster than I could respond from Tim, Dave, Mike, and others. Before responding, I wrote out a narrative of our Sunday dives, describing the artifacts and their conditions, then sent it out to all the *Kad'yak* Hunters, as we've started calling ourselves. One thing I asked was whether we should try to recover the cannon before winter storms set in. But most of the professional archaeologists in the group believed we were not ready to conserve it yet. Besides, in my opinion, it was in a pretty stable location, and I knew we could find it again. Via a flurry of back-and-forth email, we decided to leave all the remaining artifacts alone. I thought I had done all that I could do this season anyway and didn't really have any reason to return to the site, although Stefan had indicated he wanted to go back to Monk's Lagoon for some surface photography and narration.

There was one nagging concern throughout it all. Josh and Steve had gone back to the site today, on their way over to make some dives on the west side of Kodiak. Ostensibly, they were just going to visit the *Kad'yak* once more, but some folks worried they might take one step further and remove some artifacts. I had talked with Josh yesterday and specifically asked him not to disturb anything, and he had assured me we were "on the same page." I had no reason not to trust him to keep his word. Nevertheless, Dave called the Kodiak State Troopers and alerted them to keep an eye on the site. As far as he was concerned, there should be no further diving there, and I agreed with him.

For the next hour, I typed up my dive logs from the previous day, and before long, my phone was ringing off the hook. That morning, I responded to phone interviews with two Anchorage radio stations and a TV station. I stressed the fact that the *Kad'yak* discovery was the result of work by numerous people and named some of the principals involved. I also emphasized that the wreck was in state waters and under its jurisdiction, and that any looters would be in violation of state laws. The interviews took a while, but the rest of the day went relatively smoothly, and I actually finished proofreading my manuscript and sent it off to the publisher. My desk was finally reappearing from under the piles of unfinished projects.

I was determined to finish up all my works-in-progress that had been sitting on top of my desk for weeks. On Wednesday, I planned to head over to Afognak Island, where I was to be the guest scientist at Dig Afognak, a remote camp for kids that featured traditional Native arts, crafts, and sciences. In two weeks, my family and I were going to go on a kayak trip down the Stikine River with my wife's sister and her family. And then in September, we were leaving for Washington, DC, where I was going to work at NOAA headquarters in Silver Spring for four months. There were only a few days left to clear off my desk before I would be gone for four months.

At lunchtime, I drove out to Walmart to get my photos. I was relieved and maybe somewhat surprised that they turned out rather well. Of the three or four photos taken of each object, at least one of each was clear, well lit, focused, and identifiable. There was some backscatter in them caused by sediment in the water reflecting light from the flashes, but on the whole they were remarkably clear for having been taken in Kodiak waters in the middle of summer. I scanned the photos into my computer and printed out copies. I planned to send a package of materials to Tim, Jenya, and Jason, which would include my dive logs, sketches of the wreck site with bearings and distances, printouts of navigation charts showing the location in detail, and now two pages with eight photos. It was a complete presentation of the wreck and I was proud of it. I just wished I could see their faces when they received it.

With all the Kad'yak excitement, I still had other obligations and switched gears that evening to prepare for my trip to Afognak Island. Dig Afognak was a program begun by the Native Village of Afognak in 1993 to restore and enhance the archaeological and cultural history of the Alutiiq people of Afognak and Kodiak Islands. They invited archaeologists and local citizens to participate in archaeological research and discovery on Afognak Island. Eventually they added a summer camp where both Native and non-Native kids could learn about and experience traditional Native arts, crafts, and culture. Dr. Sven Haakanson, Director of the Alutiiq Museum in Kodiak, had invited me to share my knowledge of marine life and tidepool creatures with the kids at camp. For the past twelve years I had been teaching tidepool ecology at Kodiak College, a branch of the University of Alaska. I taught in a multimedia classroom and at our lab, where there was a touch tank with all the important creatures in it. But at Afognak, I would not have any of those tools.

I briefly considered taking over my laptop, a digital projector, slide projector, and all the high-tech tools I normally used. But after talking it over with my wife who was a retired high school teacher, I decided to go low-tech. For one, I didn't want to ruin any of that equipment out in the wilderness and I didn't know how steady their power supply would be. For another, I didn't want to turn off kids by boring them with slideshows. Meri had an extensive collection of children's books on marine science, so we sorted through them to see what was useful. I selected a few and went back to the lab late that night to print out a bunch of photographs for Meri to laminate.

When I returned home, I learned that I had been on television. Anchorage's Channel 13 News had run a story about the *Kad'yak* using film they had received from somebody. I was sure it was the short clip that Stefan had made for me, but I didn't know who sent it to the TV station. Once again, I was shocked to learn they had showed the anchor, the cannon, and virtually all of the wreckage. Why was this happening? Despite my best efforts to keep all that stuff out of the public eye, it was coming out of the woodwork. There was nothing I could do to stem the tide of information about the wreck. The story was just too enticing, and the bush telegraph was working seamlessly.

WEDNESDAY, JULY 30–FRIDAY, AUGUST 1, 2003

Wednesday morning, I rose at six. Cailey and I were supposed to meet a boat at eight o'clock in Anton Larsen Bay for the ride to Afognak Island. It was a forty-five-minute drive over the pass to the bay, and as we were going out the door at 7:15, we got a phone call telling us our departure would be delayed an hour. Not long after that, the phone rang again. It was the *London Times*, calling me for an interview. We talked for half an hour, and despite their requests, I refused to release photos of the cannon or anchors. I would stand my ground on that until I was ready.[7]

We finally got out the door after that. On the boat ride over to Afognak Island, several of the adults asked me about the wreck, but to tell the truth, I was tired of talking about it and kept mum. It was a relief when we arrived at the campsite on the remote island an hour later, and we quickly settled into the routine of camp. I was happy that Cailey had come to spend time with me and glad to get away from the phone and my email for three whole days. Cailey and Meri had been away for five weeks, attending music camps, history camps, and other summer vacation activities during the entire period leading up to and including the discovery of the wreck. I really missed her. The storm could swirl without me for a while.

During the next two days Cailey and I explored tidepools with groups of kids, hiked through the woods, swam in a (cold!) lake, ate camp food, played tag on the beach, made crafts, painted rocks, and generally goofed off. It was a great getaway, and we came back relaxed, refreshed, and reacquainted with each other.

AUGUST 2003

Before leaving Kodiak for our family kayaking trip, I wrote up and sent another grant pre-proposal for the *Kad'yak* search to Tim Runyan, who would be the principal investigator for the project, and Frank Cantelas, one of Tim's colleagues at ECU who would serve as the chief archaeologist. It would be more likely to get funded if an archaeologist was the PI, so I was grateful for them to take the lead on the project. I would act as the local coordinator for boat charters, dive gear, and supplies, and help schedule volunteer divers, but I really didn't expect to play a pivotal role in their survey.

Meanwhile, I would dedicate myself to learning more about the history of the *Kad'yak*. Why was the *Kad'yak* carrying ice? How did ice become such an important Alaskan export? And how could that knowledge help us understand and identify the wreck site?

CHAPTER 13

PHANTOMS
OF THE
DEEP

ICE. IT MAY SEEM HARD to believe, but ice from Woody Island was the single most valuable commodity in Russian America in the 1850s. Ice was the reason the *Kad'yak* came to Kodiak and floated into Icon Bay. Although the Russians had come to Alaska for the sea otters, by 1850 sea otter populations in Southwest Alaska had been virtually exterminated and were generating little income. The market for furs in China was beginning to collapse as a result of the Opium Wars and subsequent competition with (forced) British imports. Incursions were being made into Alaska by Hudson's Bay Company, which had set up an outpost at Fort Yukon on the Yukon River and had begun offering the Natives better prices, plus rifles, in exchange for furs. British ships had begun scouting the Alaska coastline in the late 1840s, ostensibly looking for the lost Franklin Expedition but surreptitiously trading with the Natives and undermining the Company's influence. Hostilities with England increased during the years leading up to the Crimean War, which lasted from 1854 to 1856. By this time, the RAC was bloated with managers and bureaucrats. Most of the Russians present in Russian America were retirees. Thus the greater part of the RAC budget was spent just maintaining the lives of nonworking residents and their churches, schools, and hospitals. Russian America had become a financial drain on the RAC and the Russian Empire. It's no wonder that by 1850 they began considering the sale of their American colonies to the US, although nothing was done about it at the time. Economic stagnation threatened to bankrupt the company.

That suddenly changed in 1850. The California Gold Rush created a new class of wealthy citizens in San Francisco who needed ice for refrigeration and to put in their drinks. The only viable source for ice was Boston, which meant the ice had to be shipped around Cape Horn, a long and perilous journey of several months. Even though up to half of the ice melted during this journey, it was still a profitable business. However, it soon became apparent that another source was needed, and Alaska seemed like the perfect choice. In 1851, the American-Russian Company was formed in San Francisco specifically for the purpose of purchasing and shipping ice from Alaska to California, and it immediately set up a contract with the RAC for supply and delivery. (It was also possibly a cover-up arrangement by which Russian territories could be ensured a supply of goods during the Crimean War.) Ice was initially cut from ponds in New Arkhangelsk. The first shipment of 250 tons of ice was delivered by the *Bacchus* to San Francisco in 1852 and was sold for $75 per ton, an unheard of price at the time, for a total value of $18,750. Within a year, the American-Russian Company had entered into a three-year contract with the RAC to purchase 1,000 tons of ice annually at $35 per ton.

When the *Kad'yak* was purchased in 1851, it was the last ship added to the fleet of the RAC, which then numbered ten vessels. It's highly likely that the *Kad'yak* was purchased specifically for the ice trade, as that was its primary purpose for its entire career in Alaska. Some of the ice ships had specially built elevators for handling the ice. Ice was placed on a steel tray, and as it was lowered into the hold, its weight raised another tray up to the deck level. It may be that such equipment was present on the *Kad'yak* as well and could be found on the bottom of Icon Bay.

Initially, the RAC planned to cut ice in Sitka, but warm winters made poor ice, and much of it melted during the voyage to California. By 1855 the ice-cutting operation was moved to Kodiak, which was a better source. The place they chose to make the ice was a lake on

Woody Island. Not far from the beach was a shallow lower lake and a larger upper lake farther back in the woods. The lower lake received a lot of wind and wave-born ocean spray, so it did not produce good ice. But the upper lake, Tanignak Lake, was surrounded by trees and protected from the ocean spray. It wasn't very large at first, but because of the high demand and price for ice, the Russians enlarged the lake by building a dam. The dam raised the level of the lake by 16 feet so that it had a depth of 22 feet in the center and a total area of 44 acres. Water flowing over the dam also powered a sawmill for cutting lumber that was used to build warehouses for the ice and sawdust in which to pack it. To move the heavy ice blocks from the lake to the shoreline, the Russians built a wooden flume that was set into a cut in the hillside. From there, the flume crossed the ice of the lower lake and ended at the warehouse. Eventually they built a second warehouse, and the two of them could store up to 12,000 tons of ice.

Ice cutting began in December, when ice thickness grew to 12 inches. The RAC brought Aleut Natives and some Russians over to Woody Island to do the work. First, they marked off sections of the ice. Special horse-drawn saws were used to cut the ice, after which workers would move it to the flume and chute it down to the lower lake and the warehouse. Sawdust from the mill was used to insulate the ice in the warehouse, which protected it from the often mild and rainy weather. By February, the warehouse was full, and shipments to San Francisco began in March. At its peak, the lake on Woody Island produced nearly 6,000 tons of ice per year. Ice was being shipped not only to California but to Hawaii, Mexico, and as far as South America. As a direct result of the ice business, Woody Island became unique in Alaskan history; it was the only place that horses were ever used in Alaska, since they were unable to travel anywhere else in the territory; it was also the location of the first road ever built in Alaska, which ran around the perimeter of the island, in order to exercise the horses when they were not being used to cut ice.

During the ice-cutting season, up to two hundred Natives and Russians worked at the operation. Workers were paid $0.20 per day, plus two cups of rum and a hot bowl of fish soup at noon. When Alaska was sold to the US in 1867, the ice operation was taken over by the American-Russian Company, who raised the workers' wages to $0.40 per day. In 1868, Angus McPherson, then-manager of the Woody Island ice works, died suddenly, leaving instructions for his son Jonathan to take over as the new supervisor. The Company, though, had a low opinion of the son and instead offered the job to Captain David Walker of its ship *Olga*, which often made ice runs to Kodiak. On the first day of Captain Walker's tenure as superintendent, the Aleut workers on Woody Island declared that they didn't want to work for their new supervisor and went on strike for a day. Captain Walker assumed they had been put up to this by the junior McPherson. After some discussion, Captain Walker offered to forgive them if they would go back to work the next day, and some did; but those who didn't found themselves without their meals or rum and were refused permission to purchase food at the company store. The next day, all were back at work.

The round trip from Sitka to Kodiak, to San Francisco, and back to Sitka took about four months. With fair winds, the trip from Kodiak to San Francisco took about five weeks. By 1859, the price for ice had declined to $6 a ton, and even though up to 10 percent of the cargo could melt during the trip, it was still valuable enough to make the trip worthwhile. But in 1867, Andrew Muhl invented the artificial icemaker, and the Woody Island ice business began a gradual slide into oblivion. Price wars occurred, and advertisements were placed in the San Francisco papers touting the high quality and healthy effects of using "natural" ice instead of the artificial kind. (Substitute the word "salmon" for "ice" and it sounds exactly like the arguments being used today by fishermen in the battle between farmed and wild salmon.) In addition, the opening of the transcontinental railway made it possible to ship ice down from the Sierra Nevada Mountains to San Francisco. Due to declining prices and competition, the value of Alaskan ice declined precipitously. In 1879, the artificial ice producers bought out the contract for ice production from the American-Russian Company. They were paid to harvest the ice and store it in warehouses on Woody Island, but not to ship it. Just in case the artificial icemakers failed, there would be a backup supply. But it was never needed, and in 1880, the ice production came to an end. Ice was never again shipped from Woody Island.

In the early 1860s Russia entered into discussions about selling Alaska to the US. During the discussions between Russian Ambassador Baron de Stoeckl and Secretary of State William Seward, Seward offered an opening bid of $5.5 million. This was more than de Stoeckl expected, but being a good horse trader and hoping for a fat commission for himself, he continued to bid it up. A final offer of $7 million was eventually agreed on, and de Stoeckl asked that the US also take over the contract for ice production from Woody Island. Seward would not agree to accept an existing liability, but in consolation, and without consulting Congress, he upped the price by $200,000 to a final price of $7.2 million. The men shook hands. Ice from Woody Island was so valuable and important in the end that it was arguably responsible for the sale price of Alaska being increased by about 3 percent of the total value. All this from a little lake.

The ice carried by the *Kad'yak* became a phantom cargo that disappeared after the ship sank. Over time, ownership of the ship also became a specter, lost in the shuffle of records following the sale of Alaska. Within a month of our discovery, another phantom ship loomed out of the deep, murky waters of Kodiak's history. And with it, the ownership of the *Kad'yak* came back to haunt us.

LATE AUGUST 2003

The initial euphoria of finding the *Kad'yak* site has turned sour. Two bronze rods that were initially recovered by Steve Lloyd still have not been returned and probably remain in his possession. Dave had tried to contact Steve by phone before catching his flight back to Anchorage the day after our discovery dives, but he wasn't able to reach him. About a week later, Dave tried again and sent an email to both Steve and Josh, requesting the return of the items. We were willing to give them the benefit of the doubt at first; perhaps they had just put them in a dive bag and forgotten them. We didn't want to be accusatory. But now after repeated requests, the items were still unaccounted for. I feared it was all part of a plan by Josh and Steve, hatched well before the dives ever occurred. In keeping with Admiralty Law, which preceded the Abandoned Shipwreck Act, any person may lay claim to a shipwreck under the

Law of Finds or the Law of Salvage by having the courts "arrest" the wreck until court proceedings were finished. An artifact removed from the wreck was traditionally presented to the courts as a proxy for the ship itself. Whoever holds the artifact can claim ownership of the wreck.

In 2001, Josh and Steve had used their magnetometer to locate another sunken ship. The *Aleutian* was a steamship that sank in 1929, in Larsen Bay on the west side of Kodiak Island. At the time, it was carrying several freight cars of copper ore, miners, fishermen, and various other travelers on a journey from Alaska back to the Lower 48 (an Alaskan term for the Continental United States, referring to Alaska's location on the "north-up" globe, but perhaps also encompassing its opinions of its inhabitants). Steaming into Larsen Bay at full speed of 9 knots, the *Aleutian* hit a rock and sank within ten minutes. Thanks largely to the quick action by the ship's captain, all personnel got safely into lifeboats and rowed away before the ship sank, although one unfortunate soul decided to return to the boat for some treasured item and went down with it. The ship now lay in about 210 feet of water, and the rock where it sank has since been known as Aleutian Rock. Efforts were made to locate the ship soon after its sinking, but the technology of the time was not adequate for the job. The *Aleutian* had remained unfound for seventy-four years.

The *Aleutian* is of minor importance as a historic relic. The copper ore might be worth something if it could be recovered, but the expense probably wouldn't justify it. Could there be other, more valuable items on the ship? Rumors that the ship was carrying a large amount of money or gold in the safe flew around Kodiak for a while but were never substantiated. At any rate, few took interest in finding the ship until last year.

Sometime back in 2002, Josh had mentioned to me that he and Steve had been diving on the *Aleutian* in the previous year and planned to do so again in the summer of 2003, but since I knew nothing about it, it had little meaning to me. Now, not more than three weeks after finding the *Kad'yak*, the *Anchorage Daily News* ran another front-page article announcing the finding of the *Aleutian*. The article also stated that "Lloyd and Lewis were among the divers who found the historic wreck of the *Kadi'ak* last month, also off

Kodiak. The *Kadi'ak*, which sank in 1860, is the oldest shipwreck ever discovered in Alaska waters." The last sentence was almost a verbatim copy of my statement quoted in the *Kodiak Daily Mirror* on July 23. Otherwise, there was not a mention of me or any other details about the *Kad'yak*, which was just as well.

After keeping mum for a year, why would they make this announcement now? The answer was right there: They planned to file a salvage claim on the *Aleutian*. According to the article, they hoped to attract divers to visit it as an interesting dive site. That didn't make much sense to me. There were plenty of wrecks that were both more interesting and easier to reach, in water that was much clearer and warmer, in other parts of the world. Who would want to travel all the way out to Larsen Bay just to go scuba diving in those dark, cold waters unless they lived here already? I couldn't imagine the hobby wreck diver venturing all the way up here for a few thrills, especially at a cost of $4,000. Besides, it was so deep and in such a remote part of the world that only the most skilled and experienced cold-water divers would be capable of making the dive. There must be another reason.

The article also mentioned that Josh and Steve had removed some plates from the wreck, which they were going to use when filing the salvage claim. If there were no competing claims on the wreck, the judge could issue a decision in favor of the claimants. Other competing claims might occur if an insurance company had made a payout on the ship and wanted to claim its right to salvage, or if the ship was owned by a foreign company. Or if the State of Alaska wanted to assert its right to ownership. The *Aleutian*, as it happens, was also inside state waters, and the state would definitely assert ownership.

As far as the State of Alaska was concerned, the *Aleutian* was an archaeological site just like the *Kad'yak*. And according to the article in the *ADN*, the Department of Natural Resources and the state attorney general's office were looking at the *Aleutian* events as a serious infraction of state law. Although they did not need a permit to dive on the *Aleutian*, Josh and Steve did need a permit to "survey" it with a magnetometer and another permit to remove any items from the wreck. In addition, one requirement to obtain such a permit was to have a plan in place for conservation of any artifacts recovered and a participating archaeologist on site during exploration and recovery. They had none of that.

At the time of our first *Kad'yak* dives, Dave had heard rumors of the *Aleutian* discovery but had few details. He had not met Steve until they were aboard the *Melmar* for the *Kad'yak* dives, and then he put two and two together because he knew that the owner of Title Wave Books was one of the principals of the *Aleutian* fiasco. Dave's radar was already on high alert before mine even switched on. He was not sure he could trust Steve or Josh but hoped for the best. Their plans to lead dives on the *Aleutian* were legal, as long as they did not do anything as blatant as try to recover or sell artifacts from it. But once Dave was alerted that clients would be allowed to collect artifacts as souvenirs, the State of Alaska became legally involved. Kodiak hunting guide Harry Dodge admitted to receiving a plate from the *Aleutian* wreck, but I wasn't sure who gave it to him. However, when Dave's requests for return of the *Kad'yak* artifacts were ignored, it became clear that there was more to Josh and Steve's disappearance than just oversight. Perhaps they had held on to the *Kad'yak* artifacts in order to use them as bargaining chips in their quest to wrest ownership of the *Aleutian* from the state. But the attorney general's office would have none of that. As far as the state was concerned, the two issues were independent, and it would not participate in any such bargaining.

A day or two later, I received a phone call from Jeremy Weirich, an archaeologist with the NOAA Office of Ocean Exploration in Silver Spring, Maryland. He knew Tim Runyan and had reviewed our previous proposal for the *Kad'yak* that had not been funded. Jeremy had recently talked with Tim about the missing artifacts and was concerned. He didn't know me and asked what my connection with Josh and Steve was. Was I involved with the *Aleutian* discovery and salvage claims? I wasn't sure why he was asking me these questions, but it soon became clear. The stories that had appeared in the *Anchorage Daily News* about the *Kad'yak* and the *Aleutian* had both been mostly written and orchestrated by Steve. Both stories suggested that their salvage company had been largely responsible for the discoveries. Jeremy apparently thought I was involved with the salvage company and the disappearance of the artifacts. I was dumbfounded. I assured him that I had nothing to do with either. I was operating completely within the law with total cooperation of the State of Alaska.

His interest, he said, was sparked by Tim. Jeremy knew we would be submitting a new proposal based on the discovery of the Kad'yak's location, and he would be one of the people reviewing it. He didn't think NOAA would want to support research on the shipwreck if there was any possibility that a competing private salvage claim could be filed on it. I'm sure I convinced him I had nothing to do with any such claims, but I was astonished that anybody in NOAA would even think so. I worked for NOAA. Why would I do anything to jeopardize that relationship? This was not a good turn of events.

A few days before leaving Kodiak, I saw Josh Lewis walking down the street with someone. I stopped my car to talk with him. I really wanted to ask Josh what was going on, but not in public, so I asked him to call me on the phone later that day. He promised to do so, but I didn't hear from him.

I did get a call from Stacey Becklund though, who confirmed my worst fears. According to her, Josh had come to see her at the Baranov Museum and told her point blank that he and Steve were going to file a salvage claim on the Kad'yak. When Stacey questioned this, she said Josh had told her they had known about the wreck for many years and had invited me to accompany them on their trip to examine it. He implied that the whole discovery was the result of his work and my involvement was purely secondary.

I was completely astounded. *How could Josh be so two-faced?* I wondered. Stacey's revelation supported everything I had been thinking and elevated all my nagging doubts to levels of complete certainty. Since this was hearsay, I couldn't verify any of it, but it was entirely consistent with Josh and Steve's behavior and explained everything perfectly.

I called Dave immediately. Could Josh and Steve actually make a claim on the Kad'yak? According to Dave, all they needed to do was to file a claim in federal court, using the bronze pins as evidence. Would the state intervene? If they knew about it, yes, but there was no requirement for the feds to notify the state of any claim.

If someone showed up to contest it, there was a good chance it would not be awarded, but there was no way to guarantee that the state would learn about such a filing in time to protest. Furthermore, Josh and Steve had a very savvy lawyer named Peter Hess who would be looking into insurance claims on the *Kad'yak*. (In fact, the Alaska Office of History and Archaeology was being advised in the *Aleutian* case by NOAA General Counsel Ole Varmer, who had won three previous legal battles against Peter Hess.) Dave also thought it was likely that the ship was insured by Lloyd's of London, although verification of that would require more research. If an insurance claim was filed after the sinking, ownership of the *Kad'yak* would have been transferred to Lloyd's, if it was the insurer. And it was possible that Lloyd's would give up its rights to the wreck if requested because of the time, distance, and lack of salvageability.

So what was the next step? Dave had not had any response to his informal request for return of the artifacts. The next thing he would do, he said, was to have the Alaska Department of Law send a certified letter to Steve and his attorney requesting the return of the items. If they did not respond to that, then Dave could have the state troopers physically go and remove the items. That would mean searching Steve's house, his place of business, whatever. Maybe even Josh's house and boat. But for now, Dave could only send a letter and wait.

Hearing that was disheartening. I sank lower in my chair. I needed to do something, but what? I couldn't defend my interest in the wreck unless I knew what was happening, and there was no way to find out in advance. Josh wouldn't talk to me. I tried calling him at work and at home, but he wouldn't answer or return my calls. I had to fight back somehow. I had to go on the offensive.

With great trepidation, I called the attorney general's office in Anchorage and talked with Deputy Attorney General John Baker, who had authored the letter to Steve. He told me the same thing that Dave had. They could have their staff keep an eye on the federal court agenda, but they would have to check it daily to keep from overlooking it. I asked if they could do something proactive, like file their own claim in advance, but he said there wasn't any reason to do that because the state already owned the resource. John was concerned about the wreck issue, but he admitted it was a low-priority item for them. They just wouldn't do anything until and unless a competing claim was filed. I told John I would do whatever was necessary to help prevent the *Kad'yak* from falling into private hands. Would I testify in court if necessary? he asked. Absolutely, I responded. Would I sign affidavits? Absolutely. But for now, there was nothing to do but wait and hope nothing happened.

Meanwhile, ownership of the *Kad'yak* was uncertain. The State of Alaska claimed ownership by virtue of the ship's location and the Abandoned Shipwrecks Act. There was still some possibility that it was owned by an insurance company, but which one was anybody's guess. Tim Runyan believed that ownership and salvage rights would have transferred from the RAC to the US with the sale of Alaska and all their other property. Another possibility was that ownership had been abandoned during the Russian Revolution of 1917. And yet another was that it belonged to the Alaska Commercial Company, which took over assets of both the RAC and the American-Russian Company after 1867. The worst possibility, even though remote, was that Josh and Steve could file a claim on it in federal court simply by walking in with a bronze rod from the ship. All of these possibilities drifted through my mind, like phantoms in the fog, competing for my attention and drowning out everything else.

The next morning, I woke up early and walked over to Josh's house, which was only about a block away, and knocked on the door. It was eight thirty. He wasn't home, but his wife, Vic, assured me that he knew I wanted to talk with him and that he would call me. I went back home. He didn't call.

I couldn't wait any longer after that because the very next day I was on a plane headed for Washington, DC, where I would live for the rest of the year.

CHAPTER 14

EXILE
IN THE
BELTWAY

Since 1991, I had been applying for, and receiving, research grants from the National Undersea Research Program (NURP). In that time, they had probably given me over one million dollars in research money. Of course, none of it had personally gone to me. Most of it had been for chartering use of submersibles, remotely operated vehicles, survey equipment, and vessels to operate from. Occasionally, I had gotten a few dollars for some piece of equipment, but that was rare. Most recently, I had been awarded about $100,000 each year in 2001 and 2002 for continuing research on Tanner crab spawning aggregations.

In early 2002, I began to think about a temporary assignment in Washington, DC. There were a number of reasons for doing this, and foremost was the thought that my supervisor was planning to retire soon, and spending some time at NOAA headquarters might be a nice feather in my cap when it came time to apply for his position. In addition, I wanted to learn more about the NURP and Ocean Exploration (OE) programs with the thought that maybe someday I might want to apply for a position with one of those programs. So I contacted Barbara Moore, the director of NURP, and told her of my interest in her program. But I had no specific agenda or project in mind at that time. We agreed to keep in touch and discuss more at a later time.

In June of 2002, I met with Barbara in Washington, DC, to discuss a potential temporary assignment with the NURP program. She informed me that NOAA had a program that might facilitate my working with her. It was called the NOAA Rotational Assignment Program (NRAP) and was designed primarily to offer employees at headquarters an opportunity to get some experience working at field offices for periods of three to six months. Each participating office would submit a brief job description for a project that could be conducted over that time period, and they would be advertised nationally. That program could also be used in reverse, to bring in employees from the field for experience in headquarters.

The NRAP program would require me to travel to Silver Spring, just north of the DC border, and live there for a period of time, which would be expensive. I would have to convince my supervisor or division director to cover the cost. But Barbara offered me another way. Her office needed help and was planning to submit a proposal to the NRAP program that would cover the costs of my travel and living expenses. Would I be interested in applying for that?

I wasn't sure at first. She wanted someone there as early as summer of 2002, but I was thinking more about the following year, 2004. Besides, I had other research projects that would tie me up until June, and I did not want to spend my summer in Washington, DC. The weather at that time would be muggy and hot, and I hated to leave Alaska in the summertime, the only time of year when it was really enjoyable. Furthermore, at that time I was still hoping that the NOAA OE would fund our *Kad'yak* proposal, which would keep me busy in July. But I agreed to look at her proposal and consider applying. The program would not be advertised until March 2003, and I had until April 10 to decide.

Then I learned in March of 2003 that the *Kad'yak* proposal was not going to be funded. By that time, the NRAP program was open, so I took another look at the offerings and saw that both NURP and OE had temporary positions available. I applied for both. A month later, I was surprised to get a call from the NRAP program coordinator. I had been selected by both programs and needed to make a choice. I was elated, but it was a hard decision.

Over the next few days, I learned as much as I could about both programs from their websites. The Undersea Research Program had been in existence for about fifteen years. Their main office was small, with only a handful of staff, and they had a limited website. There were five regional centers around the country, each specializing in various types of undersea research in their localities with a core group of scientists who received grant money from them. My impression of NURP was that it represented a respected, highly scientific program that funded well-planned, cutting-edge research. As the director, Barbara Moore was a respected administrator who promoted focused, hypothesis-driven research programs. She and I had had several conversations, and I thought she was very personable and straight-forward.

Ocean Exploration, by contrast, was a new program, an upstart. They had a large headquarters staff, and a big portion of their job was public outreach. As such, they had an extensive website. In 2002, I had served as the chief scientist aboard the RV *Atlantis*, operated by the Woods Hole Oceanographic Institute (WHOI), to conduct research on Gulf of Alaska seamounts with the research submersible *Alvin*, and OE had funded the cruise. I had given that cruise the name GOASEx, for Gulf of Alaska Seamount Expedition, and it had stuck. Although OE had funds to spend, they didn't own any ships, submarines, ROVs, or research equipment, so they contracted that work out to other institutions such as WHOI or universities that operated ships. The OE website concerning the GOASEx cruise had a lot of information, well over fifty pages, some of which I had written. And that was only one of the projects they had funded. They were splashy. The projects they funded were risky in that they had no guaranteed payoff, and they were all over the map. Projects ranged from seamounts to geology, to shipwrecks, to hydrothermal vents, etc., with no unifying theme. The director of OE was Captain Craig McLean, an officer in the NOAA Corps. The Corps was an organization of quasi-merchant marine officers who commanded the various NOAA ships. All were college graduates with various degrees, and their assignments usually alternated between two years at sea and four years on land. Captain McLean was young, businesslike, and enthusiastic, a real mover and shaker.

I had called several times to talk with him, but I could never catch him in his office. I had even made an appointment with his secretary to schedule a call. Even then, he had not been available.

My choice eventually boiled down to this: Both programs were doing exciting work. But would I rather work for a fellow scientist who was willing to talk with me, or a ship captain whom I could never reach on the phone?

SEPTEMBER 2003

Meri and Cailey had decided to stay in Kodiak for a week longer, so I traveled to Silver Spring by myself and spent a solitary week there. I rented an apartment in a high-rise just across East-West Highway from the NOAA headquarters buildings. In ten minutes, I could go from my apartment on the fifteenth floor, down the elevator, across the street (dodging the traffic because there was no crosswalk nearby), into the NOAA building, and up the elevator to my cubicle on the tenth floor. It was a simple commute.

The NOAA complex consisted of four buildings labeled SSMC1 through SSMC4, which stood for Silver Spring Metro Center. I was working in SSMC3, the largest building. It housed the National Marine Fisheries Service (NMFS, my permanent employer), the National Ocean Service (which conducted bathymetric surveys and produced navigation charts), Oceans and Atmospheric Research (which included NURP and OE), parts of the National Weather Service, and other parts of NOAA. Several cubicles away from me was the Office of Ocean Exploration. My first week at NOAA headquarters was spent trying to figure out where NURP fit into the big picture of NOAA and just exactly what it did.

But while my days were filled with new names, places, and experiences, my nights were filled with worry. What was happening with the *Kad'yak*? Why hadn't I heard from Tim or Dave? Did Josh have the artifacts? Were they filing a claim on the ship? How could I protect my interest in the wreck site? And what exactly were my interests? Should I file a claim on it so I could facilitate the research? No, I decided, that would be the worst thing to do; I would then be no better than any other claimant. The wreck belonged to the people of Alaska, whose ownership was protected by the state. Then did

I care that my name should be associated with its discovery or further study? No, I did what I set out to do, which was simply to find it. From here on, I didn't really care what people said or whom they thought was responsible. It was found, and that was the only important fact. Then why was any of this business troubling me?

The answer came immediately: because I wanted to see it through to completion. The wreck site needed to be surveyed completely by competent marine archaeologists. Artifacts needed to be recovered, preserved, and conserved, and they needed to be made available to the people of Alaska and Kodiak. The *Kad'yak* was their heritage and would make a significant contribution to knowledge of the history of Russian America. Tim and his colleagues were world-class scientists and had the skills and technology to do the job right. I did not want anything, least of all a battle for ownership, to interfere with that.

That same week, Tim, Frank, Dave, and I submitted a two-page pre-proposal for work on the *Kad'yak* to the OE office next door to my cubicle. Tim and Frank would be the principal investigators, and Dave and I would be co-principals. After receiving the final version from Tim and learning that it had been received at OE, I walked around the corner and introduced myself to Jeremy Weirich. His background was in archaeology, and he and Craig would conduct the initial review of all pre-proposals. It was merely a formality, he assured me, to weed out those that were inappropriate, incomplete, or just plain wacko. Apparently they got a lot of those. We briefly discussed the missing artifacts, and I suggested to him that I did not think Josh and Steve would file a salvage claim. I just couldn't see any point in it. However, I'm not sure I convinced him that I was anything other than naïve.

A week later my family arrived, and we began to settle into life in the nation's capital. Meri and Cailey spent their time trying to set up dance lessons for Cailey. She was a tenth-year ballet student and wanted to continue ballet classes, but she wanted to get some other training too, including Irish step dancing and hip-hop. She also wanted to learn a new musical instrument, the Uilleann or Irish bagpipe. Unlike the Scottish bagpipes, into which a player blows air, Uilleann pipe players must pump a bellows with the right elbow and squeeze a bag with the left elbow, all while playing the notes with the right hand. In the hands of an expert player, its sound was ethereal and much lighter than that of the Scottish pipes. Who would imagine a fourteen-year-old girl from Alaska ever desiring to learn how to play such a contraption? But she was sincere, and we managed to find not only a teacher but a set of pipes she could rent.

The first week after Meri and Cailey arrived, Hurricane Isabel blew up the coast and into Virginia. By the time it reached us, it was no worse than any storm we'd experienced in Alaska, and we had difficulty understanding why everyone else was so concerned. The DC Metro was closed Thursday afternoon, and all government employees were told to stay home on Thursday and Friday. When the winds came on Thursday night, they blew down trees all over Virginia and Maryland and knocked out power to over a million homes. We looked out our windows as the lights blinked out all over Silver Spring, including the apartment across the parking lot from ours and the grocery store next door. But our building did not lose power.

During this time, I received an unexpected email from Josh Lewis: "Send us some photos of Hurricane Isabel." I couldn't believe it. After all we've been through in the last two months and his avoidance of me for the last month, how could he send me such a chatty letter? As if nothing whatsoever had happened between us. It was baffling. But it meant that he was not averse to communication, and I knew I had to respond. Just how was the question. I did not want to burn bridges any further than they had been already, but I needed to be blunt. For a month I had been rehearsing what I would say to him if I had the chance. But I didn't think that a phone call was the right forum for the conversation, and I also didn't think I would actually get a straight answer out of him. So I'd have to send him an email. He may not answer it, but he would read it. For the next week, I thought carefully about how to say what I needed to say without being accusatory or derogatory. Finally, I just sent him a terse note saying that he owed me an explanation for his behavior. I honestly didn't think he would answer. But at least he knew what I thought.

The following week I received a short but astonishing email from Dave. It said: "Steve Lloyd's lawyer brought the *Kad'yak* artifacts to my office." Apparently, the attorney general's office had finally sent them a letter telling them they had better turn over the artifacts, or the state troopers would be knocking on their doors. It worked.

A major sigh of relief escaped my mouth. But what did it mean? We now had all the known collected artifacts in our hands. Did Josh and Steve have a change of heart about filing a salvage claim? Did my letter to Josh have anything to do with it? My previous experience with them led me to believe they had a plan behind every action, so perhaps there was more to this than met the eye. Did they collect other artifacts that we didn't know about? Was this just a smokescreen to make us think the conflict was resolved and thus drop our guard? Despite what looked like a peace treaty, I still didn't trust them. Was I wrong?

A few days later, another letter arrived from Josh. He made some statements to the effect that the *Kad'yak* artifacts should be placed in a museum in Ouzinkie (a remote village on Spruce Island of about fifty people with one store, one church, and no museum), that they would probably be removed from Kodiak by the state, and that he and Steve were trying to protect the artifacts and not deprive the people of their heritage.

Josh's letter left a lot to be read between the lines. I suspected that many of his concerns were promulgated by his friend Reed Oswalt, who had ties to Ouzinkie. Reed had previously stated that "the stories belong to Ouzinkie Village, and had been handed down for generations, because this was their heritage." Maybe so, but I didn't see that that gave them ownership of the shipwreck. Furthermore, when pressed for details, Reed couldn't recall ever hearing any of the stories himself. Certainly, the Natives of Ouzinkie should be consulted and brought into the project, and that was to be a core part of our grant-funded outreach program. In order to go ashore and survey the shoreline, both to verify some of Arkhimandritov's bearings and to scout for other artifacts that may have washed ashore over the years, we would need their permission. We also wanted to show respect for the people who have kept the legends and stories about their island alive for many years. But I doubted that placing any recovered artifacts in a museum in Ouzinkie would be a practical solution. It was too remote and difficult to reach; the average traveler would never go there. Perhaps Reed thought it would help to establish a tourist industry for Ouzinkie, but I believed that the greater benefit would be obtained by displaying the artifacts in an existing museum in Kodiak.

TUESDAY, OCTOBER 21, 2003

Earlier this month, Tim had forwarded me a copy of a letter that he had received from Craig McLean. Our pre-proposal for work on the *Kad'yak* had been approved, and we were being asked to submit a full proposal. I had breathed a big sigh of relief. The next step could go forward.

Today I said goodbye to Meri and Cailey at Dulles Airport and boarded a flight back to Kodiak, where I was scheduled to give a presentation at the annual meeting of the Kodiak Historical Society. It was to be the first time that I or anyone else would speak about the *Kad'yak* in a public forum. That evening, over a hundred people, including Stacey Becklund, filled the meeting room at the Kodiak Senior Center. Relying on the help and research of Evgenia Anichenko and Dr. Gary Stevens, a historian at Kodiak College, I led them through the history of the *Kad'yak* and talked about the Woody Island ice business. I admitted how I had struggled with the interpretation of Arkhimandritov's bearings and shared the *Ah-ha!* moment when I saw Lydia Black's copy of his map. I showed the videos of diving, provided by Stefan Quinth, and the detailed layout of the wreckage and my photos of the artifacts. Although I had some trepidation about revealing the presence of cannons and anchors, I felt now that most people probably already knew about them anyway, and they would be much more willing to support our efforts if we were transparent. I concluded by stating that finding the shipwreck was just the beginning of the process—the real work would be to survey it, and eventually recover and conserve some of the artifacts for public display.

Questions followed. Eventually someone asked the inevitable: "Wasn't there an issue with possession of some of the artifacts recovered from the ship?" I replied simply that all of them were in possession of the State of Alaska, and we had no reason to think there were any more at large. Then Stacey dropped a sandbag on me.

"Did you know that a research permit has been issued to East Carolina University?" she said. "And that Josh Lewis and Steve Lloyd have filed a salvage claim to the *Kad'yak*?"

I was flummoxed. Why would Stacey bring this up in a public forum? Why hadn't she told me about this earlier?

I replied I had not heard any of that, but I was sure that the ECU scientists were competent and professional, that this was probably part of the process, and that Josh and Steve's claim had no merit whatsoever.

And secretly I hoped her information wasn't correct. We had already invested a significant amount of time and resources in developing our own research program. But I felt as if I was being placed under a microscope. How could I presume to be the spokesperson for this project if I didn't know important information like that? I needed to get back into the information loop, and quickly.

FRIDAY, OCTOBER 24, 2003

I was in Anchorage to play concerts with the Kodiak Island Drummers when the phone rang in my hotel room. It was Dave. I asked him about Stacey's remark and if Josh and Steve had filed a claim on the wreck site. His answer was slightly comforting, but not totally satisfactory.

In order to bolster their chances on a claim, Josh and Steve needed to have a written research and conservation plan for the *Kad'yak*. For that purpose, they had contracted with an archaeologist named Stan Davis. Stan was a graduate of the Texas A&M Archaeology Department and had done some work in Alaska, but it wasn't marine archaeology. He knew Dave and had called him to ask about applying for a research permit for the *Kad'yak*. All of that was perfectly appropriate, and the two had a cordial relationship.

Luckily, Dave had already anticipated this move and had previously encouraged the ECU team to apply for a research permit for the *Kad'yak* as well. He had even helped them prepare the permit in order to insure an independent review process, and the permit had already been approved, with Frank Cantelas as the principal investigator. It would be valid for three years, with the option of a three-year renewal. During this time, the state can exclude any other parties from accessing or investigating the site. In effect, that would prevent Josh and Steve from having any further operations there. Dave explained to Stan that he could not issue a permit to another party at this point, and Stan knew the system and got the message.

There was only one standard caveat to our permit: Work has to be initiated within three months, or the permit would be voided. For that reason, Dave called to ask if I could go back to dive on the site before the end of the year.

"I can't," I said. "I'm going back to Washington tomorrow, and will be there until January." I asked if "work" could include examination of the already recovered artifacts.

Dave wasn't sure, but he said he would look into it. He could also ask Tim and Frank to reapply for another permit if the first expired too soon. Only he would know about it.

"But does that mean that Josh and Steve can't file a salvage claim?" I asked.

"No," he said, "it doesn't prevent that. They could still do that if they have artifacts that we don't know about. And they don't even have to notify the state. But if the state knew about such an application and contested it, the existence of a valid research permit would almost certainly negate any competing claim to the site."

Dave assured me he would keep tabs on the federal court filings over the next few months, but it wasn't a foolproof system. While the state asserted its claim to the wreck site under the ASA, Dave and state historian Joan Antonson had also been working behind the scenes to reinforce the claim by initiating the often lengthy process to nominate the site to the National Register of Historic Places.

After we hung up, I felt better. We were not out of the submerged woods yet, but we were getting there.

NOVEMBER 2003

I called Tim to arrange a visit to ECU. I wanted to meet him and the team there in person, and hoped to give them some photographs and charts to include in the proposal. To my surprise, he told me the full proposal was due the next day! I had expected the due date to be in January like it was last year, but this year the deadline had been moved up. How could I have overlooked it? I felt like an idiot.

Luckily, Tim and Frank were on top of it and were planning to send it out by FedEx tomorrow. After work that day, I ran home and spent two hours trying to email some photographs and charts to them. But the files were so large that they would not go through on the first few tries. Only after deleting all the stored files and trash in my mailbox could I get the files to go through. I breathed more easily after that.

The next week, my family and I drove down to Greenville, North Carolina, to the university to see Tim, Frank, Jenya, and Jason. We developed an instant rapport. I was particularly delighted to find Jenya to be such an enthusiastic and energetic person. Then I gave a talk to a class of a dozen or so graduate students and retold the story of how the ship sank, my interpretation of the bearings, and the discovery of Arkhimandritov's map. I showed them the diving videos, but the best part was showing them the photographs of the wreckage and how the artifacts were arrayed on the seafloor. I even told them about our kayak trip to Monk's Lagoon and the dreams we had there. Meri thought I had gone over the cliff with all this metaphysical stuff, but I think she secretly enjoyed seeing a part of my personality that wasn't completely scientific. After the class was dismissed, Frank, Jason, and several other students stayed and examined some of the photos. We had a great discussion about the artifacts and what they could be.

In the evening we all met for dinner at a local restaurant and shared stories about shipwrecks, ship construction, Alaskan history, and diving. We had so many interests in common that the conversation never lagged. It was as if we were best friends already. They all took to Cailey instantly, and she to them. Jenya showed up in a homemade jacket that was a replica of those worn by Russian sailors in the nineteenth century. I was impressed to learn that it had been made by another student as one of his projects! When we departed, I gave Jenya a big hug. I was so happy to be working with such a competent and friendly bunch of people. It was a good omen.

The whole trip was immersed in shipwreck preservation and excitement. The next day Tim took us to a warehouse outside of town where they were conserving artifacts from the *Queen Anne's Revenge*, the ship sailed by Blackbeard the Pirate. The ship had been scuttled in Beaufort Inlet, North Carolina. Over several years of diving, the ECU students had recovered timbers, cannons, bullets, cannonballs, and a variety of other artifacts. Cailey was fascinated by the idea of pirates and really enjoyed seeing some of those items. She kept making faces at me and saying, "Arrr!"—a trick she picked up from Tim.

The process for conserving the cannons and other large objects was extremely complex. Each cannon lay in a large vat made of wood and fiberglass, about 8 to 10 feet long and 3 feet wide. The vat was filled with sodium carbonate and had two large metal plates in it, one on each side, that were connected to a large battery charger plugged into a wall outlet. The electric charge on the plates created an electric current in the fluid, which drew off the carbonate and rust layer encrusting the cannons and replaced it with metal ions. Small bubbles rose to the surface, evidence of hydrogen gas given off during the process. Because of this, the warehouse had a huge fan at one end to create adequate ventilation. The cannons we were looking at had been in the electrolyte vats for two years and would probably have to remain there for another three years. After that, they would be dried out and coated with a polymer to keep out water. The whole process was actually pretty simple.

We could do this in Kodiak, I thought, *if we had an adequate building.*[8]

My family and I then drove to Newport News, Virginia, to visit the Maritime Museum. This past summer, the turret from the USS *Monitor*, a Civil War–era steamship, had been recovered from 250 feet in the Atlantic Ocean, and it now resided in a tank at the museum. Rather than being submerged, it was under a large shower, which kept it sprayed with water. Apparently, it was not yet ready for electrolysis, which would probably take up to ten years.

I had made arrangements to look at documents in the library, which were under lock and key and only available to qualified researchers, so once at the museum I searched through old documents and books about ship construction, trying to determine what the metal artifacts were that we had seen and recovered. From examining ships at Mystic Seaport, I had already determined that the metal rods, or drift pins, we had found were bolts used to connect the knees (supporting the deck) to the ribs of the ship. After several hours, I found some drawings that looked familiar. One showed a flat metal plate with holes for bolts in it, exactly like one of the pieces of metal we had recovered. It was part of the rudder hinge, called a gudgeon. However, I still could not find any large, cylindrical objects, like the one we had seen on the Splashcam, and later found, in any of the photos or drawings. I thought back to the object with chain links near the bower anchor. Could it be part of the capstan or anchor windlass?

With more questions than answers, we returned to Washington. I was extremely happy to have finally met the ECU team, and found them to be such friendly and likeminded folks. And I had renewed my enthusiasm about the *Kad'yak* and our future archaeology expedition.

DECEMBER 2003

On December 8, I received a cryptic email from Stacey Becklund that was so brief, I did not know what to make of it. Apparently, the lawyer Peter Hess had come to Kodiak to talk about making the *Kad'yak* into a recreational dive site. Later that afternoon, I received a phone call from a friend who read me part of a news story about Hess's arrival from the *Kodiak Daily Mirror* (see Appendix B.2). I was flabbergasted. What now? Were they still trying to make a salvage claim on it? Had they hired this lawyer to press their case?

On a hunch that the *Anchorage Daily News* might pick up the story, and that Josh would not let it go unreported, I checked their website the next day, and sure enough, there it was. Hess was hired by Josh and Steve to work for them and had given a talk at Kodiak College. He had claimed that Josh and Steve had discovered the wreck and that it should be used as the basis for a recreational dive industry. Worse yet, the story revealed they had visited the site with a Russian diver and may have disturbed some of the artifacts. I was incensed. I specifically made a point of not confirming in print the type of artifacts found, though Josh and Steve have already done so. Inviting outside divers to view it just spread the knowledge of its location.

I called Tim and we had a long talk. Should we fight fire with fire? Would it do any good? Tim's had plenty of experience with these kinds of people before. On one hand, he said, we must fight back when people challenge the right of scientists to study archaeological sites. If we don't, the public will only hear the pseudoscience, misleading arguments, and disinformation being spread by our detractors. On the other hand, recreational divers should be our best allies. Archaeologists have learned that if they exclude the diving community, it created only animosity and trouble. Instead, he has learned to involve them in his projects at the outset. Recreational divers can be persuaded to support our work and to participate in it; and if they did, they would be more willing to take "ownership" and responsibility for protecting the site from looters. Removing artifacts from sites was not in the best interest of the diving community because most of them were not preserved properly and ended up rusting away in somebody's garage, and it would then deprive later divers of the opportunity to view them in situ. Ultimately, it was unlikely that we would ever recover all of the *Kad'yak* artifacts, so Tim said one of our stated goals should be to "prepare" the site for recreational diving by preserving what artifacts we can, removing to museums those that could most easily be pilfered, and preparing a useful map to help divers study the site. The *Kad'yak* would someday become a bona fide recreational dive site. But it belonged to the state and should not end up being the property of any private owner.

I spent the next day writing a long letter to the editor of the *Kodiak Daily Mirror* (see Appendix B.3). Until now, I had refrained from talking in public about the removal of artifacts by Steve. I tried to be totally aboveboard, giving them credit for assisting in the discovery. But now I had my fill of their shenanigans, and all gloves were off. In my editorial, I described how I had determined the position of the wreck, how Josh and Steve came into the process at the last minute, and how they hid artifacts from the state and refused to return them until threatened with legal action. I explained that the *Kad'yak* would make a poor dive site and only benefit a few self-interested parties. I also pointed out that the research permit granted to ECU restricted access to the site, and the visit by Steve and Josh was in clear violation of that. My letter ended with a plea that Kodiak should be united in the process to survey, recover, and conserve the *Kad'yak*. We had a chance to set a precedent for caretaking of Alaskan shipwrecks, and we should not squander the opportunity to do it correctly.

After sending my letter off, I received an email from Adam Lesh, who was now the publisher of the *KDM*. He wanted their lawyer to review my letter in order to make sure I had not said anything that could be construed as libel. Later that day he sent another email in which he said they had concluded that I had not said anything that Steve and Josh had not already admitted publicly via the *Anchorage Daily News*, so they wouldn't have any grounds for complaint. But I didn't trust Josh or Steve, and they were both independently wealthy. I would not be surprised if they did try to sue me just to try to shut me up. Adam agreed to publish a slightly shorter version of my letter but with my statements intact. That would stir the pot, I knew.

I didn't have long to wait for a response. Within a week, Steve had penned his own letter to the *Kodiak Daily Mirror*. To describe it as an attempt at personal and professional character assassination would be generous. He denied taking the artifacts without permission and did not admit he took them off the boat. He called me "a career bureaucrat" (which I found particularly humorous) and said my claims were "...misleading, inaccurate, and...self-congratulatory grandstanding." After reading Steve's letter, I didn't know whether to be insulted or just laugh about it. At first, I thought I would not respond, that the tit-for-tat had gone far enough. Tim encouraged me to write back, to try to keep the upper hand on the subject. I wasn't sure what good it would do. But within a day, I had penned another letter, laying out all the mistruths in Steve's letter.[9] Dave McMahan also sent a letter to the *KDM* refuting much of what Hess had said and explaining Alaska's interest in preserving the wreck site. (see Appendix B.4)

Over the next week, Tim, Dave, and I had discussions by email about whether we should create a *Kad'yak* website. I was still trying to keep the photographs and site descriptions under wraps, but Steve had given out just about everything except the position of the wreck. Mike Yarborough agreed to develop the webpages. Eventually, I relented to the idea. I put all my text and photos on a CD and sent it off to Mike in Anchorage.

I had planned to work at the NURP office in Washington, DC, through Christmas week, but on Monday morning, I handed in my last project assignment to Barbara Moore and just spent the rest of the day organizing my papers and computer files. Barbara wouldn't have time to review the material until after the holidays, so there really wasn't any sense in my staying around longer. I packed up my papers and sent them back to Kodiak, said my goodbyes, and went home. Anyway, my father was visiting from California, and I wanted some time to spend with him. During the week, we alternated days between visiting local museums and packing and shipping home our household goods.

On December 29, my family and I flew to Tacoma to spend New Year's with Meri's mother. We caught serious colds that week, and by the time we arrived in Tacoma, we all had sinus infections. We drove straight from the airport to a doctor's office and got antibiotics. Maybe it was the stress of moving or maybe it was just the season, but whatever the cause, I spent most of that week in bed, coughing, wheezing, hacking, and sneezing, and taking a regular regimen of antibiotics, decongestants, vitamins, cough drops, and painkillers. But through it all, I had one thought: Get the *Kad'yak* archaeological survey back on track.

CHAPTER 15

BACK
IN THE
SADDLE

JANUARY 2004: AFTER LIVING IN another city and another home for four months, it felt good to be back in my own life again. I enjoyed returning to work and tending to my crabs in the laboratory again. It was real work, unlike the paper pushing I had done at NOAA headquarters, and I felt much more productive.

I was also much closer to the *Kad'yak* and the action around it. Early this month, Dave called to invite me to a meeting he was organizing in Anchorage to discuss issues surrounding the management of submerged cultural resources. He said his office would pay for Tim and Frank to attend, as well as Jeremy Weirich from NOAA OE. I was excited. If possible, we'd get them down to Kodiak to hold a public meeting and discussion of shipwreck management and our project with the public and other interested parties. And maybe, just maybe, we might even get the opportunity to dive on the *Kad'yak*.

Soon after, I called the board of the Maritime Museum and successfully arranged to attend one of their meetings. They had sponsored Peter Hess's talk in Kodiak, and I was concerned that they might be buying his story, that shipwrecks were available to the taker on a "finders, keepers" basis. Over the past few months I had read parts of the Abandoned Shipwrecks Act of 1988 and the Alaska

Statutes on Historical Preservation. They reinforced my opinion that Hess was way out of line with reality, and I wanted the museum board to know that as well.

FRIDAY, FEBRUARY 27, 2004

For two days at the submerged cultural resources meeting, I had talked with professional archaeologists about marine archaeology. It was a refreshing break. I felt fortunate to be included among a group of folks whose opinions I respected greatly and whose backgrounds and training were so different from mine. Mike Yarborough summed up the meeting succinctly: "I was hoping there would be a lot less talk, and a lot more beer".

I had flown to Anchorage by myself, but this morning I returned to Kodiak with a team of archaeologists, including Tim Runyan, Frank Cantelas, Dave McMahan, Evgenia Anichenko, and Jeremy Weirich. That afternoon, Stacey organized an impromptu meeting at the Baranov Museum. We invited the boards of both the Baranov and Maritime Museums, staff from the Alutiiq Museum, and both the city and borough mayors. We wanted the community to meet the ECU team and learn what we planned to do that summer. It was also important to meet with representatives from Ouzinkie. Weeks earlier I had sent letters to both the Ouzinkie Native Corporation and the Tribal Council but hadn't received a response. But at the last minute, Stacey and I made several phone calls and finally reached Paul Panamarioff, a board member of the Ouzinkie Native Corporation, who agreed to attend the meeting.

About thirty people were present, and all seven of us made brief statements about who we were and what we hoped to accomplish. Until now, I had been the only person in Kodiak whom the public associated with the *Kad'yak* besides Josh Lewis. All our public arguments were simply one against the other, so I could see why some people had trouble distinguishing which of us was telling the truth. But seeing all these professional archaeologists together and hearing their arguments about the importance of the *Kad'yak*, why it should be preserved and not given over to private enterprise, had a definite effect. And my association with this group of professional academics who obviously knew what they were doing also helped to boost my own credibility. By the end of the meeting most of the museum board members and others had come around to our point of view.

After the meeting I had a brief chat with Tim and Dave to talk about a possible dive on the *Kad'yak* the next day. The weather report looked good for February in Alaska, so we decided to go ahead with our plans. Verlin had agreed to take us over to Icon Bay. I knew that his boat would be cramped for the seven or eight people who wanted to dive, yet I didn't want to leave anyone behind. Tim, Dave, and I walked down to the harbor to look at the *Ursa Major II*, a boat owned by the Fish and Wildlife Service that I hoped we could charter for our expedition next summer. On our way back up the dock, I saw harbormaster Marty Owen on his boat, the 45-foot *Sea Breeze*, and on a whim I asked him if he would take part of the group to Icon Bay for the dive on Saturday. To my surprise, he said yes as long as we left early in the morning because he had business to take care of that afternoon. I breathed a big sigh of relief and agreed, and then hustled Evgenia off to have dinner with my family.

That evening we held another public "Shipwreck Show" at the Kodiak Senior Center. Tim gave his condensed view of the Maritime Studies Program at ECU, and Frank talked about the *Monitor*. Finally, I spoke briefly about the *Kad'yak*. It was much shorter than the talk I had given for the museum in October because I wanted to get to the high point of the evening, which for me was the question-and-answer session. That was where we would really get the community involved. The questions we received were extremely perceptive and showed that people were aware of the issues surrounding the wreck. Who really owned the site? How would Ouzinkie be involved in the project?

Most important, who was protecting the site? Although the state troopers were assigned that responsibility, the site was remote and not visible by the troopers on a regular basis. Even the Coast Guard rarely flew overhead there. And even if they did see a boat, how would they know if it was authorized or not? Ouzinkie residents and others visited Monk's Lagoon regularly, and the troopers couldn't be questioning every boat that went in there. The likelihood that boats were bringing divers was so small that it wasn't worth their time to investigate. As much as I would've liked to keep the location a secret, it was too late for that. So the best thing to do was depend upon the community to protect the site. Everyone must recognize the value of the wreck and the artifacts for learning about our history, and we must be our own watchdogs. We must all educate our friends and neighbors and prevent anyone from pilfering, damaging, or destroying the wreck.

SATURDAY, FEBRUARY 28, 2004

I was up before six this morning, ready and excited to go diving. I woke up Evgenia, who was staying with my family and me, and then we picked up Tim and drove to Verlin's shop to get our drysuits and extra weights. The other divers—Frank, Jeremy, Dave, and Pete Cummiskey, one of my NOAA colleagues—were already waiting for us on Marty's boat in the harbor. Tim and Evgenia had never worn drysuits before, so they were trying on ones that we would rent from Verlin. Soon we had all our gear and drove back down to the dock, where we loaded the weights onto the *Sea Breeze*. Tim, Evgenia, and I then headed over to the launch ramp to wait for Verlin.

A few minutes later, Verlin's boat, *Most Wanted*, pulled up to the ramp as the *Sea Breeze* pulled out of the harbor behind him. In a few minutes, we loaded Verlin's boat and pulled away from the dock. Looking up, I saw it was heavily overcast, but it wasn't particularly cold for a February morning in Alaska and there was little wind. For a while, we all stood on the top deck congratulating ourselves on our good luck with the weather. After motoring past Near Island at the "No Wake" speed, we entered the Woody Island channel, and Verlin goosed the twin diesels. Now we began to feel the ocean swells coming around the cape and could see the breakers on the reefs. What had started out as a calm morning was going to become a windy day. But after rounding a buoy, we had the wind at our back, so the ride over to Monk's Lagoon was tolerable and fairly quick.

We pulled into the lagoon and turned to face back into the wind and swell, the *Sea Breeze* arriving about fifteen minutes later. Here, the wind was blowing about 20 knots, and there were 4-foot seas rolling in. Finding the anchoring location with the GPS was a challenge in the wind and swell, and the anchor didn't want to bite at first. Finally, it took hold, and we stopped drifting and began to pitch up and down.

The motion of the boat was making me uncomfortable, and Evgenia was already seasick, so I knew it was best to get dressed and geared up as fast as possible. After checking my gear one last time, I stepped off the rail and plunged into the waves. Swimming up to the anchor line took some effort in the swell and chop, but I didn't want to use up my air and instead breathed through my snorkel. The anchor line jerked up and down above me while I held on to it, waiting for Evgenia. After about ten minutes, she finally took the plunge and swam up to join me. Then I realized that I had left the video camera behind, so I had to let go of the anchor line and let the swells push me back to the stern, where Verlin handed it to me. Once again, I swam back to the anchor line and held onto it, gritting my teeth. A month earlier I had broken my left wrist while ice skating, and for four weeks I'd been wearing a brace on it, but today I took it off in order to dive. With my right hand holding onto the camera, I was clinging to the anchor line with my weak left one, and as the swells jerked it up and down in my grasp, my wrist was starting to hurt. We hadn't even started diving yet. We hung there for another ten minutes before Tim finally entered the water.

After getting the okay from both Tim and Evgenia, I emptied my BC of air and headed down the line. I didn't get far before I noticed that neither of them were following me, so I turned around and went back up to the surface. Tim was struggling to empty his BC, which was too buoyant. While doing so, he dropped his snorkel, so I dove down a few feet to retrieve it. When he finally emptied his BC and sank below the surface, I started down the line again, looking back to make sure I was being followed. But at 40 feet, I lost sight of Jenya above me, so I turned around again and went back up for her. By this time, my 3500 psi tank was down to 2000 psi, and I was beginning to wonder if we would have to abort the dive. I learned she was having trouble with her ears, but finally she got them equalized. Once again, we headed down into the dark water. At 60 feet, I could see the bottom, so I descended some more before I released the anchor line and dropped to the sandy seafloor below me.

The visibility was about 15 feet, and I could see a reef in the distance. At first I thought I was near the west end of the wreck where there was a narrow passage between the reefs, but I didn't see any wreckage around me. Holding my compass in my left hand and the video camera in my right, I turned to face west and began to swim. In a few feet I came across what I thought was the small anchor, the kedge, we had found in July. We must have landed just about in the middle of the wreck site then. I spent a minute filming the anchor and noticed that it looked somehow different. It was larger than I remembered, and there was a large rock next to the flukes that I had not seen before. I thought that the large bower anchor was directly south of us, about 30 feet, so I pointed my compass south and swam off. In a few seconds, the bower anchor came into sight, much sooner than I had expected, but it confirmed my memory of the site. Which reminded me: I had not had time to brief the other divers on the layout of the wreck site. My group of three was now at the south end of the known wreckage, yet the others had entered the water about 50 yards south of us. They would be lucky to find any of wreck.

From the large bower anchor, we swam north to the mysterious cylindrical object. Here was the end of the guide line I had tied in July, and after a few moments we followed it north. As we swam, I saw that the line was mostly buried under sand waves that stood 8 to 12 inches tall. I didn't recall them being there last summer and could not see much of the line until it led us to the kedge anchor, where it was tied to one fluke. But it didn't look like the anchor we had just landed on. We kept going until we reached the cannon. From there we followed the guide line east to the main mass of wreckage. At this point, Evgenia waved her pressure gauge in front of me. She was down to 500 psi. It was time to head up, but I wanted Tim to have a look around, so I motioned to her to hold on for a bit. I knew she was concerned, but I also knew that it was enough air for the ascent. For the next minute I ran the video camera over the wreckage, then finally tapped Tim on the shoulder and motioned him to the surface.

Unfamiliar with the drysuit, Evgenia couldn't slow down as she headed to the surface. Even I surfaced faster than I should have, though I was dumping air from my suit as fast as I could with one weak hand and a video camera in the other. We popped to the surface about 20 yards north of Verlin's boat, and the wind and swell helped us reach it quickly. Getting out of the water was a struggle in the choppy waves, but once we were up, we were elated. We had seen the entire wreck site as I knew it. The weather was now behaving like it really was February in Alaska, and the ride back to town was rough and uncomfortable, but the cold beer we shared at Henry's afterwards was well worth the trouble.

SUNDAY, FEBRUARY 29, 2004

Tim, Frank, and I met at Harborside to define what we would need for a full archaeology project. We would have to include recovering the most significant artifacts, conserving them locally, and paying for all of the work associated with that. The total cost came out to about $1.3 million, more than twice what I had estimated. We continued working at my office and completed a two-page request for the money that would be submitted to Senator Ted Stevens' office for an appropriation request by the Baranov Museum. We also looked briefly at the video I had taken underwater the previous day. It was only eight minutes long, but it showed virtually all of the known wreckage, so I was very happy with it.

Later that evening, Tim, Frank, Dave, and I drove over to visit Gary Edwards. Gary had shown up at our Friday afternoon meeting and was lobbying hard for us to use his crab fishing vessel, the *Big Valley*, for next summer's dives. I had chartered the *Big Valley* many times before for my crab research, but I thought his 98-foot boat was too large and expensive for us to use for this project. Ever the salesman, Gary feted us with wine, cheese, and fresh strawberries, and offered use of his boat at a bargain-basement price of $1,500 per day, $1,000 lower than his usual fee. I liked Gary a lot, and it was a tempting offer, but we were still short of the money needed for the boat charter, so I didn't see how we could do it.

OVER THE NEXT FEW DAYS, I had several opportunities to examine the videotape I had shot during the dive on Saturday. I concluded that there were at least three anchors present, and the first anchor in the video, where we landed, was not one we had previously seen. It was larger than the kedge anchor and more similar to the bower anchor in shape, though smaller. And there was a large rock next to this new anchor that wasn't present in my photo of the kedge anchor I had taken in July. The new anchor was also a lot closer to the large bower anchor than I remembered the kedge anchor to be. I remembered that when we swam north from the cylindrical object, we passed over the kedge anchor that was tied into my guide line but mostly buried beneath the sand. I concluded that this new anchor must be an auxiliary anchor, for use when the bower anchor was too big and the kedge too small.

I also concluded that the sand had shifted around a lot since our dives in July of 2003. The sand wavelets had not been present last July, when the bottom there was mostly cobble. Some things, like the kedge anchor, had become partially reburied while other items not seen before became uncovered. Frank also pointed out boards (deck or hull?) underneath the large aggregate, which we now knew was the main ballast pile. I had not seen that during our dives in July, and when reviewing a photo I took of that concretion, I could see why—they had been buried under a layer of gravel and sand. Apparently it had been scoured away during the winter, exposing the wood beams. That also explained why that mass of stuff was undercut; it lay near a narrow passage in the reef that probably funneled water during the storms and tides, alternately burying and unburying the timbers.

Reviewing the videos rekindled the fire inside me. What else lay underneath the sand? I couldn't wait to find out.

MARCH 2004

The front pages of the Kodiak news reflected my excitement about our progress last month. People in the community were aware of what we were planning to do, and almost everyone was supportive of it. But we still had one nagging problem: We didn't have enough money to launch the expedition without a major infusion of cash. Specifically, we needed money for the boat charter.

This month Stacey wrote up and submitted a request for $25,000 to the Rasmuson Foundation in the name of the Baranov Museum, based largely on the request that Tim and Frank had submitted to Senator Stevens' office. She also submitted a smaller request for $5,000 to Conoco-Phillips, one of the largest oil companies in Alaska. Meanwhile, in North Carolina, Frank contacted some people he knew in the Arctic Science Program, which was part of the National

Science Foundation (NSF), and on my end I contacted the NURP office in Fairbanks and submitted a request for rapid-response funds. My initial conversations with Ray Highsmith, the director of the program, were positive.

Several weeks later, I got a call from his office telling me they might have some funds available for us. It was just what we needed. But no sooner had I finished telling Stacey the good news on the phone, I received another call from Ray. He told me they had made an error in their budget and that they actually couldn't offer us any funding. Emotionally, I felt like I had been from the top to the bottom of a rollercoaster in less than twenty-four hours.

With no money for the project in sight, I refocused my efforts, concentrating on my crab research, and tried to put the *Kad'yak* out of my mind. But it just wouldn't go away.

MAY 2004

Stacey received a letter from the Rasmuson Foundation stating that her grant request had been turned down. We were both terribly disappointed. Back in February, Dave, Tim, Frank, and I had met with a staff member from the foundation when we were in Anchorage, and she had been very positive toward our efforts. We had felt assured then that we would qualify for funding from them, so we were somewhat puzzled that our request had been denied. Diane Kaplan, the Rasmuson Foundation's president, was coming to Kodiak for a meeting later that week though, so Stacey and I decided to keep the news to ourselves until I had a chance to talk with her.

When I finally was able to sit with Diane for a few minutes, I learned that the reason we had been turned down was that we had applied for a Tier 1 grant, which was supposed to be dedicated solely for capital projects, instead of a Tier 2 grant, which could be used for other purposes. We had not been aware of the distinction even after our discussion with the staff member, though the grants were explained on their website. In short, we had screwed up. It was too late to resubmit the grant for this year.

Out of the four possible funding sources to whom we had submitted grant requests, two had now turned us down. Things were not looking good. The third request was for only $5,000, which was not enough for the boat charter, but it would still be a significant and useful amount of money if awarded. Our best bet now rested on the outcome of Frank's request to NSF. In my book, that was a real long shot. What we had heard from Frank though was encouraging. NSF had made inquiries through a subcontractor about the *Big Valley*, its charter costs, insurance, and many other details that normally would not be requested until a contract was being established. So we forged ahead, planning the expedition as if we had the funds already in place.

Late at night as I tossed and turned thinking about the project, I worried about funding. I felt as if I should be out there in the public eye, drumming up support for the project. Plenty of other organizations raised money in this town doing just that. My case was slightly different though. I didn't represent the museum, I couldn't legally raise public money for a federally funded project, and I wasn't even a principal investigator on the grants that had been submitted or the one funded by Ocean Exploration. As far as the granting organizations were concerned, I was just one participant who helped to find the wreck, if they read that far down into the body of the text.

If we didn't get the funding from NSF, then there would be only two choices left. One was to go around town with a hat in hand and ask for donations from local businesses and service organizations. I was reasonably certain that would bring in some money, but I wasn't sure how to go about it in a manner that would be seen as politically and legally correct. The other solution would be to ask local charter fishing boat operators to donate some time to the project as a tax write-off. I had already talked with a number of them and knew July was their busiest season. None wanted to give up any days on which they might have customers booked, though I knew that many of them did not have full boats most of the time. I hoped, though, that with enough notice, they might be able to donate a day or two to us. If we could convince eight or ten boats to do that, we could probably get most of the divers to the site, though we might be changing boats every day. It would require extra work to move all our gear around every day and would probably be a major headache scheduling the boats, but it could be done, and it would work. And truthfully, it could have a very positive benefit on public relations by involving more people in the project. But it was a last resort, and I hoped it would not come to that.

As MAY WORE ON, THE weather got warmer, the hillsides grew greener, and wildflowers began to appear. I began getting so many calls every week to go on school field trips or give tours of the laboratory to visiting students that it became increasingly difficult to get my own work done. One of my favorite events of the spring was going over to Woody Island for tidepooling. Several of the elementary schools had developed an annual outdoor education program for their fifth graders to spend three days in cabins on the island and learn about wildlife and outdoor survival. Local experts would be brought over each day to teach the kids about wild plants, birds, canoeing, survival skills, building shelters, and other outdoor skills. Usually several other biologists also went along, including some from my lab and some from the University of Alaska laboratory next to ours.

This year we had excellent weather. We spent the whole morning in the tidepools with the kids, turning over rocks, finding sleepy limpets, leathery chitons, wormlike sea cucumbers, delicate nudibranchs, grumpy hermit crabs, predatory snails, slippery, squiggly fish, and other interesting intertidal critters. As always, the highlight was finding octopuses. On this day, my group of twelve kids found five of them. Most were small, no bigger than my thumb, and the kids chased them around my bucket, generally tormenting them until the little octopuses squirted out blobs of ink and curled up in a ball. At that point I stepped in to give the poor cephalopods a needed respite from their tormenters. Sometimes we brought interesting animals back to the laboratory, but we already had a medium-sized octopus on display, so at the end of the day, I released all the little octopuses back into the water. After lunch, we biologists said goodbye to the kids and walked down to the shore, where we climbed into a skiff for the ride back to town.

By two o'clock I was back in my office, but the sun, wind, and early departure had taken their toll, and I was quite tired. I knew I would be worthless for the rest of the day, so I planned just to check my email before going home. There, I found a short message from Frank: The NSF grant had been funded. Hallelujah! Finally, the funds were complete. We were going to do archaeology on the *Kad'yak* this summer. I felt as if a great burden had been lifted from my shoulders.

I wanted to know how much money they offered and what else was included beyond the boat charter, but Frank's letter had been terse. I sent him a congratulatory reply and could hardly wait until next week for the details. Immediately, I called Stacey, who had received the same email, and we agreed not to spread the word to the community until we knew more of the details about the award. This was the first time NSF had ever given a grant in Kodiak, and I knew it would be worthy of a news article in the *Daily Mirror*. Stacey also wanted to send out a mass email to the board members, museum patrons, and other local folks who were supporting the project morally, emotionally, financially, or otherwise. At last, we were on the way.

A week later, after finally learning the budget details from Frank, I sent a press release to the *Kodiak Daily Mirror* announcing that we had received a grant from NSF to complete this year's work on the *Kad'yak*. A few days later, I talked with a reporter on the phone. The resulting article appeared on the front page on Thursday, but it was greatly abbreviated and included little more than a few quotes from our conversation. Nonetheless, it served to keep the *Kad'yak* and our expedition in the public eye, and keep the focus on archaeology instead of salvage. But I knew the controversy wouldn't just go away.

CHAPTER 16

SHIPS
AND
SUBTERFUGE

LATE IN MAY, THE *ALEUTIAN* case came up in federal district court. The State of Alaska was protesting Steve and Josh's claim for ownership of the *Aleutian*. Despite all the legal wrangling, the Department of Natural Resources still did not know the exact location of the *Aleutian* wreck, only that it was somewhere near Aleutian Rock in the upper end of Uyak Bay. Although it was certain that the wreck's location was inside state waters, the Department wanted better information. As part of its effort to preserve the site from exploitation, the state wanted to nominate the wreck for the National Historic Register, as it had done for the *Kad'yak*. But in order to do that, it needed a precise position.

Conveniently, Dave attended the proceedings briefly and learned that Steve, Josh, and Peter Hess were heading to Kodiak to dive on the *Aleutian* with some clients.

FRIDAY, JUNE 4, 2004

On the first day of June, Dave had called about a surreptitious mission. Would I be willing to fly out to Uyak Bay? He suspected Josh and Steve were taking the *Melmar* out to dive on the *Aleutian* site soon and wanted me to see if I could spot them. I told Dave I would try to arrange for a flight on Thursday. Tuesday evening

I walked down to the dock and found the *Melmar* still there, with a large tote on deck (a plastic box normally used to hold a ton of fish) and no dive gear. Yesterday at noon, I checked it once again. The tote was gone, and the deck was partially loaded with dive gear. That evening I found the *Melmar* was no longer at the dock.

Dave and I figured that Josh and company would wait until slack tide to dive on the wreck. I fired up my laptop navigation program and plotted the course they would have to take. The distance from Kodiak to Aleutian Rock was 94 miles by sea, which would take them anywhere from twelve to sixteen hours, depending on weather conditions. Assuming they had left mid-afternoon yesterday, they should arrive about ten in the morning today. Accordingly, I arranged for a plane to fly me out to Uyak Bay in the afternoon so we would arrive about the time of the tide change.

So this afternoon I found myself in a very old and very small Piper Cub at the Kodiak Municipal Airport. The craft consisted of little more than a frame made of tubing covered with airplane cloth. I sat directly behind the pilot but was afraid to lean against the bulkhead for fear of breaking through it, and watched as we ascended rapidly over the town of Kodiak and circled around to the southwest. Small patches of blue sky appeared overhead, but a thick wall of clouds was moving in from the southeast.

As we flew, I looked down with awe at the snow-covered mountains. Each ridge was followed by a deep valley, green at the bottom, with a meandering gravelly stream flowing off into the misty distance to the west. It was wilderness to the max. The rugged ground offered few opportunities to land a plane of this type without floats. If we had problems out here, we would be in big trouble. But the pilot was experienced and had flown this plane around Kodiak for many years, so I could only try to relax and enjoy the view. After an hour, we flew through a narrow pass into Uyak Bay, between sharp, craggy peaks spotted with dozens of mountain goats, looking like fuzzy cotton balls on the rocky slopes. Winds from the east buffeted us briskly as we descended down into the narrow fjordlike valley.

Comparing our surroundings to a chart I had brought, I could see we had entered the very south end of the bay and were now flying north toward the mouth. Ahead was Amook Island, and at its very southern tip was Aleutian Rock, marked by a tripod and navigation light. But the *Melmar* was nowhere in sight. We flew in a circle around the rock, looking for buoys that might mark the location of the *Aleutian*, but none were visible. There was an old cannery about a mile off to the west though, and I thought it might be where the group was lodging. There also seemed to be a boat at the dock. I asked the pilot to fly in that direction. As we got closer, I saw it was the *Melmar* and there were people on the deck unloading gear. They must have just arrived after beating into the wind for twenty-four hours or perhaps stopping for the night in some sheltered cove. Even though we didn't catch them at the dive site, finding them there confirmed all of our guesses about their trip.

Not wanting to appear suspicious, we flew off along the coastline, weaving in and out along the shore as if we were flightseeing, bear watching, or on some other business. After a few minutes, we flew back over a ridge to the east and headed north back to town. The clouds had moved further to the west and looked more menacing, but we stayed just parallel to the front for the rest of the trip. The last thing we wanted was to fly into a cloud where we couldn't see the mountain peaks, which could be disastrous. Fortunately, we arrived back at Kodiak and landed before the cloud layer descended on us. Upon landing, I asked the pilot to keep my mission confidential. I went home feeling like a character in a spy novel.

Even though Josh, Steve, and Peter were fighting the state over possession of the *Aleutian*, they had actually invited Dave and me along to come dive on the wreck. Perhaps they felt it would reduce any animosity that might exist, but after all they have done to date, neither Dave nor I wanted to have any involvement with them. We felt there was nothing positive that could come out of it, and it most likely would be turned against us somehow anyway.

So I kept quiet, but it didn't matter in the end. The State of Alaska eventually confirmed the wreck's location with the assistance of a local fishing guide, who had noted the wreck on his depth sounder.

IN THE MEANTIME, I WAS super excited about our own upcoming dive expedition. I sent out notices to all the local divers who helped us or indicated their willingness to do so, asking for dates they would be available to dive. Peter Cummiskey, our lab divemaster, assembled a list of documentation that each diver must provide to meet NOAA standards. He also arranged for a temporary lifting of some NOAA dive restrictions for this project, which would make it easier for non-NOAA divers to participate.

Stefan, who had documented the discovery dives on the *Kad'yak*, returned to town this week, excited to continue filming our work this year. It was with some trepidation that I told him there might be a glitch in his plans. Stefan had been hired by Shoreline Adventures, Steve and Josh's company, to produce a film about the *Aleutian*. They had planned to use it for promotional purposes. I had talked with Stefan about it and concluded that it was only work to him; he had no personal stake in the outcome of their business, and he did not condone their behavior in regards to the *Kad'yak*. I was satisfied that he was acting in a professional, objective manner. Tim and Frank, though, did not know Stefan, and they were concerned that his work on the *Aleutian* was a conflict of interest. They were reluctant to allow Stefan to continue filming our work on the *Kad'yak*.

I had been somewhat disappointed to learn of their concerns and had written to them in support of Stefan. I trusted him and felt that he would do a good job for us and would not let his involvement in the *Aleutian* project interfere with his role as a filmmaker. When Stefan returned to Kodiak, I explained all of this to him and asked him to contact Tim and Frank for a discussion. Perhaps they could work something out.

THURSDAY, JUNE 10, 2004

On the front page of the *Anchorage Daily News* today was a story about the court proceedings over the *Aleutian*. The writer, Sheila Toomey, had written the story on the *Kad'yak* last year, using mostly information provided by Steve. In today's article, she wrote about how Shoreline Adventures had advertised their claim to the ship and that a federal judge had awarded custody of the ship to them after the state had not responded to their claim. She didn't mention that the claim was a small-print notice in the back of the *Anchorage Daily News* and that the state had not been notified directly of the claim by Shoreline

Adventures or by the federal court. (In fact, there is no law requiring such notification. It's a nonsensical disconnect in the process of filing a marine salvage claim; but it's the law, and the state would have to deal with it.)

She explained that a week after the initial decision, the attorney general's office filed a motion to intervene, requesting a reversal of the judgment, leading to the court proceedings in May. In Toomey's story, she suggested that the state was interfering in a private business matter and that such salvage claims followed a history of over a thousand years. She didn't mention the Abandoned Shipwrecks Act of 1988, that it superseded those laws in waters within 3 miles of the coast. She also didn't interview Dave McMahan or anyone in the Department of Natural Resources about the case.

I should've been glad to see the story in the *ADN* because too often such proceedings occur behind closed doors, and the public is rarely made aware of them. But the tone of her article and the selective interviews made the state appear to be the "bad guys" in this situation. I couldn't let it lie like that. That morning, I typed up a five-hundred-word response to her article and sent it to her, requesting that it be published as an editorial. Two days later she replied to me with a letter of similar length, defending her views and calling my criticisms "snippy." What did I expect? However, at the suggestion of her editor, I shortened the letter and submitted it as a letter to the editor.

In my critique, I pointed out how the state was put at a disadvantage by the lack of notification. Steve Lloyd had argued that the state did not have the money or interest in salvaging the wreck and that he and his partners had already spent over $100,000 on research and diving expenses. To an uninformed reader, it might seem to be a compelling argument, but the same could be said about every shipwreck in the world. Those arguments had been used by every amateur treasure hunter, shipwreck salvage diver, and archaeological artifact collector about every historic site ever found. Alaska was littered with historical sites of Native villages, most of which were occupied for a few years before being abandoned by their seminomadic occupants. Yet those sites had revealed much about the ancient history of Alaska through artifacts, house sites, and other structures unearthed by professional archaeologists. In Kodiak alone, such excavations had revealed a seven-thousand-year history of occupation. The Alutiiq Museum housed a collection containing thousands of artifacts recovered from such sites.

In Alaska, it's illegal for amateurs to unearth, excavate, or remove artifacts from historic sites. Even professionals couldn't do so without a permit. Imagine if it were legal for anybody who might find such a historical site to take ancient tools, eating utensils, stone lamps, bones, baskets, or skulls and sell them on the open market. Opportunities for scientific discovery, learning, and public education would be lost forever. But that was exactly what was happening with the *Aleutian*. If recreational divers were allowed to pillage artifacts from the ship without a permit from the state or a plan for archaeological preservation, it would set back the cause of professional archaeology in Alaska for decades.

Upon arriving home that afternoon, I was amused to find the exact same story about the *Aleutian* on the front page of the *Kodiak Daily Mirror*. The story had been distributed over the AP Wire Service, so it really shouldn't have been any surprise. I immediately sent copies of my letter to the *ADN*, the *KDM*, and KMXT public radio.

It was no great coincidence that simultaneous with these events in Alaska, Dr. Bob Ballard was revisiting the *Titanic*, the great steamship lying on the floor of the Atlantic Ocean, 500 miles east of Newfoundland and 2.5 miles underwater. His concern, and the reason for his visit, was to assess damage to the wreck caused by salvage operations, tourists, filmmakers, and others who had visited the tragic site and removed over six thousand artifacts during the eighteen years since Ballard discovered it. Not only were rust and decay taking their toll on the ship, but visits to the ship by submarines had caused additional damage to the hull and superstructure. The ship's bell, and even the crow's nest, from which lookout Frederick Fleet had shouted "Iceberg, dead ahead," had been removed by salvors.

After discovering the wreck in 1985, Ballard had known that its location in international waters would make it prone to looting and salvage operations. Accordingly, he had made two fateful decisions: First, he would not tell anyone the location of the wreck. Second, he would make a strong public plea that the wreck should remain undisturbed as a memorial to all the passengers and crew who lost their lives on that fateful night of April 14, 1912. The only other thing he could have done to preserve it would have been to file a salvage claim on it for himself or for the Woods Hole Oceanographic Institute, for whom he was working at the time. He knew, of course, that such would not be appropriate behavior for a professional oceanographer and would put him in conflict with the US Navy, who had partially funded his expedition. So he had gambled that the world would not find the ship, and that if they did, it would be left alone. Unfortunately, he lost.

It didn't take long for the location to become public. Perhaps that information came from someone who had been part of the expedition. Regardless, the *Titanic's* general position, give or take a few miles, was already public record, and it wouldn't take much effort for anyone with a ship and some sonar equipment to find it. Ultimately, a company called RMS Titanic, Inc., filed a salvage claim on the ship and was awarded ownership of it. Now, for the right price, almost anybody can buy a piece of the ship.

The current *Titanic* expedition was jointly funded by NOAA, National Geographic, and WHOI. Accompanying Dr. Ballard and representing the NOAA Office of Ocean Exploration were Captain Craig McLean and Lieutenant Jeremy Weirich, who had dived with us on the *Kad'yak* in February. Ballard's goal on this trip was to determine the condition of the ship and make it public in the hopes that it would help convince the coastal nations of the world that something needed to be done. Toward the end of the expedition, on June 9, Ballard phoned President George W. Bush and other world leaders at the G-8 summit meeting at Sea Island, Georgia, to express his concern and plead with them to establish protections for shipwrecks in international waters. According to Ballard, shipwrecks were the "Pyramids of the deep." He said:

"To me, it's not different than the pyramids of Egypt. There's more history in the deep sea than all of the museums of the world combined. Yet there is no law covering the vast majority of shipwrecks, and a great deal is at risk.... We're just going into those museums, and the question is: Are we going to plunder them or appreciate them? And the jury's still out."

THURSDAY, JUNE 17, 2004

But in Anchorage for the *Aleutian* case, the jury—or rather, the judge—was in. The front page of today's *KDM* announced that Judge Harry Branson dismissed the State of Alaska's motion to intervene in the *Aleutian* case on the grounds that it had not been filed soon enough. Steve and Josh officially had ownership of the wreck site. Peter Hess called it a victory, though to us it was a great disappointment. If anyone can claim a shipwreck simply with a surreptitious advertisement, then what was the use of the Abandoned Shipwrecks Act? The judge did not allow the state to make its case based on the ASA, nor did he allow Shoreline Adventures to argue that the ship had never been legally abandoned. To Hess, that was the heart of the case. He argued that a ship was not legally abandoned unless the owner wrote a legal letter of abandonment. It's an upside-down definition, because the real world doesn't work that way. Things become abandoned when they are no longer useful, become destroyed, or the owners die or disband. No one in that situation would take the time to sit down and write a letter stating they were abandoning their property. They'd just leave it. But according to Hess and Steve Lloyd, the *Aleutian* was still owned by its insurer, so it was not abandoned. But if so, what business do they have to pillage it? Although this argument was never heard in court, it has direct bearing on the ownership of the *Kad'yak*.

Unlike the *Aleutian*, the *Kad'yak* was much more clearly abandoned because all property of the Russian-American Company in Alaska was transferred to the United States in 1867, and the RAC itself was disbanded during the Russian Revolution. Clearly, there was no letter of abandonment, but just as clearly, there was no extant owner, except for the United States government. Soon after the sale of Alaska to the United States in 1867, the physical assets of the RAC were purchased by Hutchinson, Kohl, and Company of San Francisco, which later became the Alaska Commercial Company (ACC), which had operated in Alaska until a few years ago. Steve said that the assets purchased by the ACC included the *Kad'yak*, but this was a specious argument because the *Kad'yak* was lost prior to the sale of RAC assets, and the state's claim under the ASA and recent inclusion on the National Register of Historic Places make this argument moot. As an analogy, when I was a grad student living in Seattle, I had owned an old Ford van that lost a bumper in a collision (not my fault, officer!); I had replaced that and the muffler, pitching the old parts along with a worn set of tires in the town dump. Eventually I had sold the van in Norfolk, Virginia. If Steve's argument played out, the new owner of my van would also own all the previous property I had lost or discarded, even though they were in a landfill half a continent away.

The most disturbing aspect of Judge Branson's decision on the *Aleutian* case was its potential impact on the *Kad'yak*. If Shoreline Adventures can claim the *Aleutian*, what's to stop them, or anyone else, from claiming the *Kad'yak* or any other historic shipwreck in Alaskan waters? At one time, well before locating the *Kad'yak* in July of 2003, I had the fleeting thought that if I ever found it, I should personally file a claim on the *Kad'yak* so that it would be preserved from looters. I had known that we had the technology to find the wreck, even without a permit or permission from the State of Alaska, and that perhaps it would be easier to do that and then get the permit later. It was a good thing I did not do either of those because then I would have been no different from any other wreck hunter, regardless of my intentions.

Eventually, the state appealed Judge Branson's ruling to the Ninth Circuit Court of Appeals, and a settlement was reached that was reasonably favorable to the state. One of the factors that shaped the agreement was knowledge that the *Aleutian* retained an unknown but probably large quantity of bunker fuel. It seemed that regulations surrounding HazMat releases were better defined than those pertaining to historic resources, and the state reminded Hess and his clients that Shoreline Adventures would be responsible for mitigating any release of petroleum that might occur while they were on site, even if they did not directly cause it. The last major remediation effort of a ship of similar depth, in California waters, had cost around $11 million, and the *Aleutian* was much more remote. Perhaps fighting over a few historic artifacts was not worth the risk.

CHAPTER 17

WETSUIT/DRYSUIT

TUESDAY, JUNE 22, 2004: THERE was a lot of preparation to do for the dives in July. Frank Cantelas would be in charge of all the ECU divers, but Pete Cummiskey would be supervising the NOAA divers and all the unpaid volunteers, including Bill Donaldson, Mark Blakeslee, Verlin Pherson, Stefan Quinth, and a few others. For that reason, NMFS wanted to ensure that volunteers, or "observers," had at least minimum qualifications and some recent dives to similar depths. All the volunteers needed to fill out some forms in order to dive with us, including a diving resume or history and a medical history form. This week, I delivered the forms to all the local divers and shepherded them through the process of filling them out so we could forward them to the NOAA Diving office for approval. Typical of government forms, they weren't easy to fill out and stymied a few of the divers who didn't understand them at first.

I also planned to test out my new drysuit from the NOAA diving program. After fourteen years of use, my old crushed neoprene suit from DUI was leaking like a sieve. I had patched it in all the obvious places (the seams, the folds, the zipper edges) with neoprene glue, but it was now leaking through the fabric. Twice in the last year I had sent it back to DUI for repair, and they had returned it with large areas of glue on the inside. But it still leaked. During our dives on the

Kad'yak last summer, I had borrowed a Viking suit from Verlin, but I didn't want to do that again this year. I was hoping that we would get enough money with the grants to purchase at least one new drysuit and some other gear, but in the end, we had to cut all those purchases out of the budget just to make the project happen. So rather than have the lab shell out several grand for a new suit, I requested one from the NOAA diving program. After three phone calls and two emails, I finally received an Abyss suit. It was also crushed neoprene, with a zipper across the shoulder like the Viking, but it was so buoyant that I needed at least 40 pounds of lead to get underwater.

Today I made a dive with Pete to check out the suit. We were going to dive in Trident Basin, a semi-enclosed bay immediately behind our laboratory, to the wreck site of the *Minnie B*, a fishing boat without any historical importance. Prior to getting into the suit, I put on long johns followed by a complete set of fleece-lined outerwear, including slippers. Stepping into the suit, I pulled it up to my waist then reached down to ensure that the leg portions came all the way up to my crotch. *Never can get into one of these without breaking off at least one fingernail*, I thought as I unrolled the rest of the suit up my torso. In order to get my head through the neck seal, I reached behind me and grabbed the seal from the top, then stretched it down over my head, nose, and ears. Finally I pulled the zipper across my chest and snugged it up tightly.

We put the rest of our gear into the back of the pickup truck and drove down the driveway to the pump house. There, we put on our weight belts. If I were diving with a wetsuit, I would wear about 20 pounds of weight on a belt around my waist. But a drysuit traps much more air inside it, so I wore 35 pounds of lead shot in a belt that hung from my shoulders by suspenders. This took the weight off my hips and lower back. Next I wrapped on ankle weights that weighed 2 pounds each to help keep my feet down (and head up) by preventing air from entering my booties. Laden with lead, we carefully climbed down a rocky slope onto the beach while carrying our heavy tanks and BCs.

As I was putting my tank on, a clamp on my fifteen-year-old BC inflator hose broke loose. While I was examining it, I tested another clamp on the other end of the hose, and it, too, broke loose. *Come on,*

I thought. Pete went back up to the lab to get some plastic cable ties to secure my hose while I sat on the beach. It was a warm day, and I was sweating something fierce. After about fifteen minutes, Pete returned and we fixed my hoses. It was a welcome relief from the heat when we stepped into the cold water.

From the pump house, we followed the laboratory water intake lines down to the 50-foot depth. That line ran out 700 feet to a steel box where water was drawn into the laboratory. Another parallel line ran 1,600 feet out to the deepest part of Trident Basin, at 80 feet. As we swam along the pipeline, we picked up interesting creatures for display in our public aquarium and placed them in mesh bags. I collected several sponge-encrusted hermit crabs and a rare red sea urchin.

About halfway out the pipe, we came to a small buoy floating above the pipeline. Pete had placed this marker there before to help us locate the sunken boat nearby. From there, we followed our compass needles south for 50 feet. It had been raining every day this month, and the water was cloudy with silt. Even though I couldn't see the boat, I knew it was near because it cast a dark shadow over me as I swam closer. Finally, the hull appeared out of the gloom.

The *Minnie B* was a 60-foot fishing boat that had sunk about six years earlier. Pete had visited it several times and removed all the salvageable items he could, including the wood steering wheel, brass compass binnacle, and exterior light fixtures. It was still very much intact, and we swam around it several times as he looked for other interesting items. At one point I found a Mr. Coffee appliance and held it up for Pete to see. *How's that for an artifact*, I thought. On the second pass, I picked up a heavy metal ovoid that appeared to be part of a scupper or fairlead. For a brief moment, I wondered whether it was worth salvaging. But if I kept it, I would be a "wreck hunter," and I wanted to be above suspicion. So I laid it down. I was getting low on air but continued to follow Pete as he swam around the boat one more time. What was he looking for? Eventually, he found the scupper and put it in his mesh goodie bag. It probably weighed 10 pounds and was too heavy to carry easily. As I fretted about my air, he calmly took out a lift bag, attached it to his mesh bag, and blew some air into it. That lifted the weight up off bottom just enough for him to tow it along with him. At that point, I signaled to him that I was low on air, and we turned to swim back to shore.

As we came into shallow water, I became very buoyant, and no matter how much I dumped air out of my suit's exhaust valve, I was having difficulty staying on the bottom. At 20 feet, I gave up the struggle and floated upward toward the surface, splaying myself out in a spread-eagle position to slow down my ascent rate and trying to dump air with my right hand. My ascent was faster than I desired, though not dangerously so. But when we start diving on the *Kad'yak*, we would be coming up from 80 feet at the end of each dive. Unless we had an anchor rope close by to follow up, we would be doing free ascents. Doing that from 80 feet is a lot more difficult than from 20 feet. I hoped I would get a better handle on this new suit after a couple more dives.

After climbing up the bank and doffing my tank and weights, I mentioned to Pete that I had previously picked up the metal ring and laid it back down in a different place. "Oh," he said, in recognition. "I found that thing on a previous dive and laid it down so I could find it again when I had a lift bag with me. But when I looked for it, it wasn't where I had left it, so I just kept swimming around until I found it again." Oops.

So my underwater dithering about the ethics of looting the wreck was all for naught then, since Pete had already "claimed" the *Minnie B* and everything on it. Knowing the history of the wreck, and its lack of value, he had no qualms whatsoever about removing items from it. I felt like I had trespassed on his property. That's what I get from hanging out with archaeologists—everything looks like an important artifact, even if it isn't.

THURSDAY, JULY 1, 2004

We finally got a bit of good news today: Dave's nomination of the *Kad'yak* to the National Register of Historic Places was approved. The wreck was listed due to "its association with significant events" and "its potential to yield important information." That was another point in our favor, and a setback to Josh and Steve's plan to claim ownership of the *Kad'yak*. It felt good to be winning again.

CHAPTER 18

SHIVER
ME
TIMBERS

Sunday, July 11, 2004: Finally, the day had come to begin the archaeological exploration of the *Kad'yak*. I've been looking forward to this for years. I was as excited as a kid on Christmas morning and couldn't wait to share my discovery with the ECU gang.

After an obligatory stop at Harborside, we assembled aboard the *Big Valley* in St. Herman's Harbor. The first hour was spent stowing the skiffs and receiving the safety talk from skipper Gary Edwards. Everyone on board needed to know the location of the life raft, fire extinguishers, EPIRB, radios, and how to use them. If the vessel had been chartered by NOAA, this would be a required event; regardless, it was standard practice aboard Gary's boat. In addition to the skipper and myself, the dive crew included Tim, Frank, Dave, Evgenia, her husband Jason, ECU's divemaster Steve Sellers, and NOAA diver and expedition photographer Tane Casserly. Rounding out the crew were the deckhand Bryce Kidd and his wife Jesse, who was also to be our cook. By nine we were out of the harbor heading toward Icon Bay. The weather was incredible. No wind, flat seas, and 75°F—it was about the best you could ask for in Kodiak.

Because of my work as a marine and fisheries scientist, I spent a lot of time at sea aboard fishing vessels, probably totaling more than four hundred days and nights in the past twenty years. The

Participants in the 2004 *Kad'yak* expedition. Top, left: Steve Sellers, Frank Cantelas, Dave McMahan, and Jenya Anichenko listen as I brief them on the layout of the wreck site. (*Tane Casserley/NOAA*.) Top, right: Frank, Steve, and Tim Runyan enjoy a meal in the galley of the *Big Valley*. Bottom, left: Jenya and Jason Rogers study their notes in the salon of the *Big Valley*. Bottom, right: Jenya and Gary Edwards share a moment aboard the *Big Valley*.

Big Valley was like many of these vessels, but it was also unique. It was originally built in 1968 for the Gulf of Mexico shrimp fishery, probably somewhere in either Louisiana or Alabama. With an overall length of only 81 feet, a draft of 12 feet, and displacement of 169 gross tons, it was small for a Bering Sea crab fishing boat. It had been extended to almost 100 feet by the addition of a stern ramp, allowing it to function as a stern trawler, but Gary covered that up with a steel plate when crab fishing. Most of the newer boats (those built since 1980) exceeded 130 feet, and some were as long as 180 feet.

The *Big Valley* had undergone numerous refits and modifications throughout its career. At some point it had been brought to Alaska to participate in the crab fishery, and Gary had found and purchased it in Seward in the late 1980s before bringing it to Kodiak. Like most Gulf shrimpers, the *Big Valley* had been built with only a single-story deckhouse with the wheelhouse up front, where there was minimal visibility over the bow. Gary had added a second-story wheelhouse and converted the original into a pantry for storage of food, supplies, and survival suits. He had also added a "weatherbreak," a steel bulkhead that ran along the port side of the deck, from the house aft toward the stern. This was used to protect the crew from weather and waves when they were handling crab pots from the starboard side of the boat. All of these changes made the vessel more useful for crabbing. Nonetheless, as a thirty-five-year-old boat, it was starting to show its age through the rust that was continuously being chipped and repainted.

Inside the boat was a long narrow passage that led past the dining table on the portside and two cabins on the starboard side to the galley and head (the ship's bathroom) near the bow. From there, a steep ladder went up to the wheelhouse. An eclectic collection of photographs—including sepia-tinted images of African and South Pacific explorers with their native guides and some large dead beast, such as a lion or elephant—hung around the dining table. There was also a photograph of actress Barbara Stanwyck, the protagonist of the 1960s television show for which the boat was named. Opposite the dining table was a large bookshelf with a wide variety of books. Most fishing boats that I have worked on were extremely utilitarian, their libraries consisting of little more than pulp novels, repair manuals, and the not-very-well-hidden issues of *Playboy* and other "men's" magazines. Gary banned the latter type of literature from his boat because of the many women crew and scientists that he worked with. The *Big Valley's* bookshelf was instead stocked with an incredible diversity of literature, including books on art, philosophy, and psychology. On one trip, I had spent half a day reading a book by Carl Jung that I picked off Gary's bookshelf.

Not knowing the lay of the bottom, Gary was cautious taking the *Big Valley* into Icon Bay and finally anchored in 60 feet of water, about 300 feet away from the wreck site. Before diving, Dave, Frank, and I went out in an aluminum Lund skiff and dropped a weight with a buoy as close to the wreckage as we could get two GPS units to agree on. Then we went back to the *Big Valley* to suit up. It was so warm, I decided not to wear my long underwear and opted for shorts and a T-shirt under my one-piece insulated dive underwear.

For the first dive, Frank and I went in an Achilles inflatable boat, and Tane and Jenya went in the Lund. We tied off to the anchor, geared up, and plunged in. By the time I hit the water, I was frying inside my drysuit. It wasn't till then that I discovered that I had only 2100 psi in my tank, instead of the 3000 I should have, and my regulator was leaking, so it would be a short dive for me.

When all four of us were in the water, I dove under, kicked downward, and sank to the bottom at 80 feet, adjusting my ears as I went. I could see the reef a few feet in front of me and turned around to see the ballast pile about 8 feet away. The buoy line was in a perfect position. Frank had not seen the ballast pile when he dove with us in February, so we spent the first five minutes examining it. The deck timbers were still exposed, perhaps even more than before. Where they came to an end, I could see more timbers beneath them running perpendicular in what I thought was the axis of the ship. They must've been hull timbers, beneath the ribs. Numerous bronze drift pins stuck up out of the wood and the sand.

After a few minutes, Frank and I swam off along the guide line I had left last summer toward other piles of material and the cannon. The string, which the archaeologists dubbed "the spiderweb," had broken off there, so I headed south, following my compass. Visibility was better than 20 feet, and after a few minutes Frank pointed out what we believed was the windlass in the gloom. As we swam toward it, we crossed the kedge anchor, which Frank briefly examined. From the windlass, I took a compass bearing to the southwest, and swam 30 feet to the large bower anchor, which was still adorned by the same lonely sea anemone that I had seen there last year. Frank inflated a small lift bag to mark the position of the anchor and tied it to the fluke. Meanwhile, I swam around looking for the auxiliary anchor and found it about 10 feet away to the west, pointing directly at the bower anchor

Then it was time to go up. I tried to make a safety stop, but my new drysuit was too buoyant, so after a minute I popped up and found myself about 50 feet from the skiff. It was a short swim back. After climbing into the Achilles and doffing our gear, we clambered over into the Lund and then motored back to Frank's lift bag. We dropped a heavy anchor weight with a large buoy attached to it,

Diving in Icon Bay. Left: Me standing on the diving platform at the stern of the *Big Valley*. Right: Jason Rogers helps Tane Casserley prepare for a dive.

marking the south end of the site. We learned that Jenya had trouble adjusting her ears, so she and Tane did not have much bottom time and had spent most of it exploring the ballast pile.

After the first dives, we decided that clunking around in two small skiffs was too cumbersome, and getting in and out of them from the stern ramp with all our heavy gear was difficult at best and could be dangerous in rough water. I recalled that the reef top was at 40 feet here, deep enough for the *Big Valley* to anchor over it, so Gary backed up very carefully nearer to the buoy markers and dropped a stern anchor onto the reef. That held us in place and prevented the big boat from swinging around.

Most of the wreckage was strung out in a dog-leg shape, north from the anchors to the cannon, then east to the ballast pile, but the buoys we placed at either end were closer together than we had expected when viewed at the surface. We dedicated the next set of dives to finding the distance from one end to the other. After lunch, Tim dove with Jason, and Steve with Dave. Tim and Jason tied a tape measure to the base of the bower anchor, which seemed to be at the western end of the wreckage. They were supposed to stretch it out toward the ballast pile, but they became disoriented and swam around for their whole dive without finding anything.

Back aboard the *Big Valley*, it was time for lunch. To say it was cramped would be a generous description of the *Big Valley*'s galley. There wasn't room for all ten of us to sit at once, so we had to eat in shifts. Jesse put on a sumptuous hot meal that warmed us all up after our dives in the 50-degree water. We needed at least two hours to blow off the nitrogen we had accumulated at 80 feet, and the *Big Valley* was well situated for that. Gary had recently built an addition on the top deck that he called "the salon." Part of it was his cabin, but the larger outer room contained a leather couch, padded seats around two sides, and a cast iron wood stove. I have worked on many fishing boats and research vessels in my career, but I had never seen one with a fireplace. Gary stuffed a few logs in and stoked the fire, and the room was soon a steamy refuge. Divers stretched out on the couch, the seats, and the floor, and we were all soon "sawing logs" in comfortable repose. After an hour's snooze, we arose from our slumbers, climbed down the gangway to the deck, and slipped back into our cold, wet dive suits.

Gary had made a number of modifications to the *Big Valley* to accommodate the expedition and divers. On the back deck, he had placed a stand-alone structure we called "the doghouse," which included a room with four bunks and a workroom with a table for organizing, assembling, and cleaning all the cameras and video equipment brought by Tane, Stefan, and myself. Behind the doghouse, Gary and Bryce had rigged up a hot-water shower so that the divers could rinse themselves and their gear after diving, and warm up from the chilly Pacific Ocean water. To make entry and exit easier for the divers, a platform hung off the end of the *Big Valley* just below the waterline. Gary had also welded up a set of steps that went down his stern ramp so that the divers could walk down to the platform. From there, all we had to do was step into the water and swim to the buoy line before sinking to the bottom. Getting back out of the water was a little harder though, especially if there was any swell. Timing the rise and fall of the big boats stern in the swell, I had to wait for it to come down to my level, then

The *Big Valley*, a 90-foot Bering Sea crab vessel, served as dive tender and headquarters for the 2004 *Kad'yak* expedition. *(Tane Casserley/NOAA.)*

got my knee up on the platform and held on to the railing as the next swell lifted the boat, the platform, and myself with 50 pounds of dive gear up into the air.

Frank and I made a second dive around four thirty, to measure the extent of the wreckage. Near the bower anchor, Frank pounded a length of rebar into the sand and attached one end of a 100-foot tape measure to it. I picked up the tape reel and swam off toward the northeast, feeding out the tape. We swam over the windlass, then across open sand and cobble without seeing any more artifacts until we eventually located the ballast pile, where our tape ran out. We covered the reel with a rock, then swam around for a look. Thinking that perhaps some of the wreckage was up on the east reef, I cruised along its edge, where I saw several large metal elbow-shaped objects that looked like ship's knees. Tim later told me that some ships from the mid-1800s did have iron knees.

Then, less than 6 feet from the buoy weight, I found another cannon! Now we could confirm there were two. This one was within sight of the ballast pile and just at the base of the reef. I must have swum by or over it several times before without recognizing what it was. It was slightly larger than the first cannon, but the last 4 to 6 inches of the muzzle were broken off and lying next to it.

Frank and I then swam to the north reef and followed its base to the west, seeing a few isolated pieces of metal. After a few minutes we turned south and swam over sand and cobble until we returned to our starting point, having made a complete circle around the site. To anyone who has not dived in North Pacific waters, or those used to diving in the clear, sunlit Caribbean, it may seem like a simple task to swim a few yards this way or that and find our way back to the start. But when diving in 80 feet of water, which is dark to begin with, and with visibility of only 10 to 15 feet, it is extremely difficult to find landmarks, identify them, or distinguish one rock from another. It requires constant use of a compass and creation of a mental map of all the objects in sight. So I was happy that my "mental map" of the *Kad'yak* was good enough to know where we were among the wreckage. I led us back toward the east reef, where the *Big Valley*'s stern anchor was sitting on the rock, and followed the reef edge north toward the ballast pile. Along the way, we saw numerous kelp greenlings and one medium-sized lingcod, both of which make for tasty eating. *I should bring a speargun next time*, I thought.

As we climbed aboard the stern ramp of the *Big Valley*, I felt a sense of accomplishment. We had seen all the known artifacts in one dive, plus many we had not found before. I couldn't have predicted such a great dive.

MONDAY, JULY 12, 2004

We got underway a bit late today. Before going down to the dock where the *Big Valley* was tied up, I had to stop at the lab to pick up a projector and screen; we were planning to go to Ouzinkie afterward to meet with people from the village and show them what we were doing. But as soon as possible we were out on the sea, this time with our special visitor for the day Mary Monroe, my neighbor and influential board member of the Kodiak Historical Society and Baranov Museum. The weather was spectacular again this morning, flat and calm but cooler than yesterday.

As soon as we arrived in Icon Bay, Frank divvied up tasks

among the team. Tane and Dave dove on the anchors to photograph them and make video recordings. Jason and Jenya spent their dive measuring and drawing the bower anchor. Frank and Steve dropped down two iron tripods Bryce had welded together, and then dove to put them in place and string a permanent baseline between them. They would be used as anchor points. Tim and I went in the water to videotape this process and also the locations of some of the recently found artifacts, including the newfound cannon. Unfortunately, my video camera got dropped on the deck when it was being handed down, and when I hit the bottom I couldn't get any video on the monitor, so I didn't know if it was working. Then the lights went out. I carried the camera around in case it was working anyway.

After I showed Tim the cannon, we went to look at the tripods. Then we followed the baseline to find Frank, though we only found the measuring tape reel he had left on the bottom when it jammed. When we surfaced, I took the video camera apart. Not only had the internal battery been knocked loose, but a capstan in the tape-winding mechanism had broken off and the external light battery pod had leaked as well. Murphy won that round.

For lunch we had fresh king crab legs donated by a friend of Gary's. In the afternoon, I helped Jason measure the auxiliary anchor—by which I mean I just lay on the bottom and shivered, keeping him company while he did the real work of archaeologists. Sometimes he would give me the "dumb end" of the tape (with the 0) to hold while he stretched it out to measure the flukes. But for most of the dive, he just lay prone in the water and drew on a dive slate. Before surfacing, we measured the distance and bearings to the windlass, cannon, and another large metal object. All measurements were within 10 percent of my eyeball estimates made the previous year.

After all the divers were aboard, we pulled up the anchors and motored around to Ouzinkie, where we had arranged to hold a community meeting to discuss the *Kad'yak* project. We ate dinner on the way, and arrived at the dock at around six in the evening. We introduced ourselves to the handful of people who had come, and talked about what we were doing, and then we had a good discussion with them about their history and their interest in the *Kad'yak*.

Nick Pestrikoff, President of Ouzinkie Native Corporation, was most helpful and gave us a short tour of the church and local history afterwards. He told us how Captain Arkhimandritov had sent an icon of Saint Nicholas to the village for display in their church. Jenya thought this meant Arkhimandritov had felt some remorse about his lack of devotion to the saint, that perhaps he had made his donation to assuage his conscience for the Kad'yak's sinking. Nick remembered seeing the icon in the church as a child, but then years ago the icon suddenly disappeared. He hinted though that he thought it had just been removed by a previous priest who had ministered at the church some decades earlier.

Though it was a small turnout, we left around nine that night with a new appreciation for local history, feeling like we had made new friends and allies.

TUESDAY, JULY 13, 2004

I stopped at the lab to pick up some minor items this morning but was on board the boat by seven thirty. Dave wasn't there; he had left the project for a few days to orchestrate another Alaska shipwreck investigation, that of the SS Portland, with a PBS crew in tow.[10]

Jason and Jenya examine an Icon of an Orthodox Saint on a tree along the forest trail to St. Herman's church.

Frank and Steve Sellers were also missing because they had gone off to exchange air tanks, and they did not return until about nine. Then Gary moved the boat over to the fuel dock for water and took off to hunt down some chain to add to his stern anchor. In the end, we didn't get off the dock until ten thirty and arrived at the wreck site at noon. We were worried about the weather because the wind was blowing about 20 knots, but by the time we reached Icon Bay, the waves were only about 2 feet, less than the previous evening, so it was okay to dive.

Yesterday we had placed metal tripods and a baseline on the bottom. Today our job was to place them into their final positions. For our first dive, Frank and I positioned the north tripod near the ballast pile. Less than 2 feet away, we found a piece of brass hardware that looked like a wall bracket of some kind, and I drew a picture of it on my slate. We then swam to the south tripod and moved it to its final position, near the bower anchor, after which we returned to the north tripod and tightened up the baseline. We followed the baseline back to the south tripod again to check its tension, which looked good.

On our way back to the anchor line, we found something new. It was a wooden cylinder sticking up out of the sand, with two brass bands around it and a deep square hole cut into one end. What could this mysterious object be? But since we were low on air, we just made a mental note to investigate it later, and continued on our way back to the ship.

After lunch, I buddied with Jason and dove onto the ballast pile. The ballast area part of the ship was now quartered; the baseline divided it into north and south sections, and another crossline placed by Frank divided it into east and west. Jason and I worked in the southeast quadrant. While I held a measuring tape in place beneath the baseline, Jason measured the width of the framing timbers and drew them on his slate, recording their positions along the measuring tape. The frames were in pairs, each timber being 8 inches wide, with 3 to 4 inches between each pair. I didn't have much to do except shiver while watching him work in his wetsuit. Jason and Tim had both decided they did not like working in drysuits, so they dove in their wetsuits. Although this made them feel more comfortable, they used more air because they were cold. After twenty minutes, Jason ran out of air, so we ascended. I still had 1400 psi left in my tank. I was beginning to feel like an archaeologist-in-training, because I was getting on-the-job experience working with the ECU team. Next time, I would need an independent job so I can keep warm and be more productive.

On the trip back to town, Jesse prepared a delicious dinner of Pad Thai, curried chicken, and salad. It was heavenly. I didn't eat this well at home.

WEDNESDAY, JULY 14, 2004

Pete Cummiskey and Stefan Quinth joined us today. As the NMFS Lab Dive Safety Officer (DSO), Pete has to be present when we worked with volunteer divers, so he came aboard most days. Stefan was here to continue his documentation from last year, after assuring Tim and Frank that he was purely a documentarian and had no connection with Josh Lewis, Steve Lloyd, or the *Aleutian* shipwreck fiasco.

The day started overcast and completely calm, though there was a slight breeze of 10 to 15 knots by the time we reached Icon Bay. The small waves did not hinder us at all. Frank must have been happy with my amateur archeology skills from the previous days because he assigned me an independent task. My job today was to determine the exact position of the three anchors and the windlass by triangulating them from the baseline, a method used by sailors to triangulate their position from shoreline landmarks. We would stretch a measuring tape from the object to markers at three separate points on the baseline. If the points were separated by 30- to 90-degree angles, those measurements should allow us to draw lines that intersect at the position of the object.

Pete dove with me today. We descended along the north buoy line so I could take him on a tour of the site and familiarize him with the positions of the cannons, ballast pile, and timbers. Then we swam along the baseline until we reached the kedge anchor, which was close to the line. In order to take direct measurements from objects at ground level, we had to drop a weighted vertical line, or plumb line, from the baseline, which was suspended about 3 feet above the sand.

Archaeologists at work. Top, left: Measuring the bower anchor. Top, right: Jason Rogers and I measure wooden frames of the *Kad'yak*. Note the centerline above our heads. Bottom, left: The windlass, with a large anemone growing on it. Bottom, right: The first cannon found in 2003, with guideline string left by me. (*Tane Casserley/NOAA.*)

The plumb line consisted of a lead weight with a foam float attached by 4 feet of twine, and I adjusted its position until the float was next to the 80-foot mark on the baseline. With Pete holding the end of the measuring tape at the tip of the shank of the anchor, I stretched out the tape until it reached the plumb line and recorded the distance on my plastic slate. I repeated the sequence at the 70-foot mark, then at the 60. Pete then moved the tape measure to the crown of the anchor, and I repeated the three measurements again, at the 60-, 70-, then 80-foot markers. It was all fairly easy because the anchor was only about 8 feet from the baseline, and Pete and I could see each other and communicate via hand signals during the entire operation.

Our next job was to triangulate the large cylindrical object, which we have designated as the windlass. It was 25 feet from the baseline so could not be seen from there, since the visibility had decreased to only about 12 feet today. Again, Pete held the end of the measuring tape at one end of the windlass while I swam out of sight and took measurements, this time moving the plumb weight 30 feet between each measurement, starting with the 50-foot marker. I swam back to the capstan, had Pete reposition the tape at the opposite end of it, swam back to the baseline, and repeated my serial recording of the distances to the three markers. All that swimming back and forth caused me to burn up my air quickly, and I was soon down to 500 psi. I reeled in the measuring tape as I swam up toward the buoy line, passing Pete at some distance and signaling him to ascend. After our three-minute safety stop, I had barely 100 psi left in my tank. I cut that dive a little too close.

After lunch, Pete and I went in again to measure the distances from the bower and auxiliary anchors. I moved the plumb bobs to the 10-, 40-, and 70-foot markers, and took measurements from the base of the kedge anchor. Swimming back to the anchor to have Pete move the tape was an inefficient method and a waste of my air, so I indicated to him that he should move from point to point on the anchors while I was at each of the markers. But as he moved, the tape snagged on a rock. Pete and I met in the middle as we tried to unsnag it, and then I got turned around. Was he at the crown of the anchor or the shank? I wasn't sure. I motioned for him to stay at one point while I swam back and measured to the plumb bobs. During that dive, we accidentally triangulated the base of one anchor and the tip of another, but did not get both points on either one of the anchors. It can be difficult to complete underwater a simple task that would take only a few minutes at the surface. Small problems are compounded by poor visibility, cold water, and the inability to talk. With little time left, we decided we would have to complete the other two points tomorrow.

Keeping meticulous notes is an important part of the job. After returning to the boat each day, rinsing off my gear, and changing back into my jeans and T-shirt, I would take my slate into the galley and transfer all my notes and marks onto my waterproof notebook. Each page contained notes on the dives, a picture showing the shape or layout of the objects we had been recording, and a table of measurements from the centerline for each item that we measured. Today my notes included a sketch of the three anchors and the baseline, with lines drawn from various points on the line to different parts of the anchors. Later, I planned to photocopy the

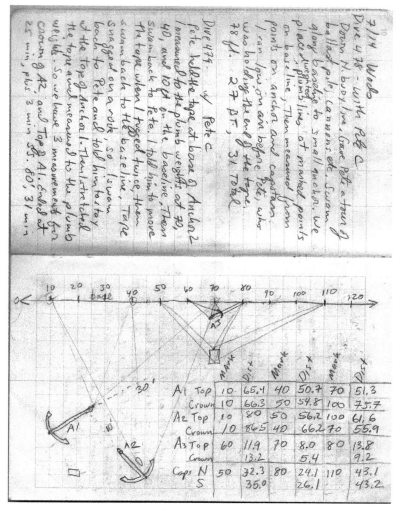

My notes and sketch of the *Kad'yak* anchors measured during dives made on July 14, 2004.

pages from my notebook and give them to Frank.

After dinner this evening, as the feelings of fullness and exhaustion, compounded by several glasses of Cabernet Sauvignon, and the warmth of the galley were dragging us into insensibility, Frank made a startling announcement.

"By the way," he said in an offhand manner, "I took a close look at the mysterious object today." I immediately knew what he was talking about. Although we had seen the wood object with brass, we were more concerned about exploring and mapping the major features of the wreck site and so had all swum by or over it. Frank continued, "It has writing on it. In fact, it has the name of the ship on it—*Kad'yak*, written in Russian. I don't know Russian very well, but that's what I think it says."

I just about choked on my dinner. "You're kidding," I said. "Did you read the whole word? Are you sure that's what it said?" His calm demeanor just wasn't what I would have expected if his story was true. And why did he wait three hours after the dive to tell us?

"No, I'm not," he replied. "I don't read Russian, but it clearly has the letters K, O, D, and some others I can't read. But I assume that's what it says."

For a second we were all speechless—before we peppered him with questions: How was it written? Where was it written? Why would they write that on an unidentifiable object? Could the word imply that it was simply an object made in Kodiak? We concluded that the Russians did not have the capacity to engrave brass fittings in Kodiak, so whatever it was, it must have been made specifically for the ship and be something important. Aside from being painted on the bow, the ship's name would usually have been engraved on the bell, and the bell was indeed the Holy Grail of shipwreck researchers. Because the wooden structure and any painted name usually disappeared after years underwater, finding the bell was about the only way researchers have of determining the exact identity of the ship. According to Tim, only one in a hundred wooden shipwrecks were ever identified with any certainty. If this object did indeed have the word "Kadyak" or "Kodiak" on it, it would verify the ship's identity. It would be an astounding discovery.

We immediately set about making plans to recover it. Tomorrow evening we had scheduled a public meeting in Kodiak to talk about the project, and having that object to display would be an explosive addition.

LIEUTENANT LARRY MUSARRA WAS A retired pilot of a US Coast

Guard helicopter. For much of his career he was stationed in Kodiak and Michigan. During that time he flew missions to rescue fishermen from disabled vessels in the Gulf of Alaska, Bering Sea, and Great Lakes. Sometimes his aircraft and crew were in as much danger as that of the seamen he was trying to help. Although he never had to ditch a helicopter in the ocean, like the events made famous in *The Perfect Storm*, he did have his share of close calls.

On one particularly stormy evening, Larry was called to rescue the crew of a fishing boat floundering in 30-foot seas near Ugak Island, off the east side of Kodiak Island. The boat had sunk while its crew managed to escape in a life raft and were picked up by skipper Joe Spicciani on the nearby fishing vessel *Ocean Hope*. The helicopter crew had been relieved that their work was over and they could return to base.

Then they received a radio call from the sinking vessel's skipper, who was still on board. He could not close the zipper to his survival suit and so had not entered the water with his crew. As darkness fell, the helicopter crew lowered a rescue basket down to the boat, now mostly awash and rolling in the waves. Larry tried to maintain a constant height above the water by dropping the helicopter between the waves, then rising up over the crest of 30-foot breakers. He remembered looking out his windshield to see water rising above his pilot's seat. He summed up the conditions and intensity of the situation in a single word—"gnarly." A large wave washed over the boat, knocking the man into the ocean. After several tries, Larry managed to drag the rescue basket underneath the floating fisherman and scoop him up into it. As they hoisted up the basket, the rotors of the helicopter hit something sticking up from the ship. Was it the steel mast? Or just a fiberglass radio antenna? Whatever it was, the rotor could be seriously damaged and might fly apart at any moment, dumping the helicopter and its crew into the rolling sea. Larry radioed his situation to Air Base Kodiak. Could he set the aircraft down on land somewhere close? The closest point of land was a high bluff at Narrow Cape. But in the wind and darkness, that was a risk too. After consulting with his dispatcher, Larry placed his bets that the rotor was stable and decided to fly back to the base. After a harrowing thirty-minute ride, the helicopter and all its occupants arrived, shaken but unscathed.

Larry was my brother-in-law, married to my wife's sister. He was retired from the Coast Guard and lived in Juneau, where he worked for the US Forest Service as the Chief Ranger and Director of the Mendenhall Glacier Visitor Center. Since he was also an experienced scuba diver, I had invited him to come to Kodiak and participate in the *Kad'yak* dives. I wanted to share the experience with him and thought it was an honor to dive with him.

THURSDAY, JULY 15, 2004

The plan today was to finish triangulating the anchors. Larry joined us today to team up with Pete and me for the task. We were all friends, having shared many kayaking, fishing, and diving adventures during Larry's days in Kodiak.

Aboard the *Big Valley*, I briefed him on the procedure, and we worked out a plan: Larry would anchor the tape, I would take the measurements, and Pete would position himself halfway between us so he could act as messenger and help clear snags when we moved the tape. Once underwater, we discovered the system worked well. Pete could just barely see each of us on opposite sides of him and could tell when one of us moved. After taking the measurements at three different markers on the baseline, I tugged the tape, signaling Larry to move his end to the other anchor. With the three of us, we finished the job with half a tank of air left.

I had promised Larry a tour of the site if we finished early, so we swam off toward the ballast pile where Jenya, Tane, and Jason were busily measuring and drawing timbers. Pete was low on air then, so I led them to where I thought the buoy line was supposed to be. But I couldn't find it. It had moved. I swam around briefly looking for it but didn't have enough air to continue, so I signaled them both to ascend. On the way up, we encountered the line arcing across our view and made our safety stops, using it for orientation.

The buoy lines were obviously out of place now. Increasing tides, wind, and waves had dragged them both away from their original positions. Going down, it wasn't too hard to determine where you were once you hit the bottom and could see a guide line or some artifact. But when our air was low and it was time to ascend, the lines were difficult to find, and wasting precious air looking for them could be dangerous.

Meanwhile, Frank, Tane, and Steve recovered the mysterious artifact we kept passing over before. They carefully triangulated its position from the baseline and photographed it in place. Then they lifted it out of the sand, wrapped it in a towel, and placed it in a 5-gallon bucket suspended below a lift bag. After bringing it to the surface, they hooked it to the ship's crane, and Gary hoisted it on board. It was indeed a wooden cylinder, about 8 inches in diameter and a foot long. One end had a square hole about 3 inches wide. An iron shaft of some type must have been there once, but it had long since deteriorated. The other end was covered with a brass end cap flanged on the outer edge with a raised center. There, engraved in the metal, were the Cyrillic letters К О D Я К ы—Kod'yak—Я being the Cyrillic phoneme "ya". Jenya explained the spellings of *Kad'yak* and *Kod'yak* were interchangeable and that the final symbol was a pronunciation sign that hadn't been used in that context since the late 1800s.

The wood itself was heavily degraded and looked as though it had been part of a larger item. What item of this shape and size would have been so important as to have the ship's name engraved on it? The more I thought about it, the more it pointed to one thing—the ship's wheel. Was this the hub of the wheel? The wood was flared out at the ends, as though tapering to a narrow edge. There could have been spokes radiating out from it that were now decomposed. We couldn't know for sure what it was, and perhaps we never will.

But it clearly had the ship's name on it. Rarely do marine archaeologists find such a telltale item on a historic shipwreck. On steel ships of the twentieth and even late nineteenth century, the name was welded to the bow, but on wooden ships, little if anything would survive to identify the ship. In fact, the ECU team had been working on the *Queen Anne's Revenge* in Beaufort Inlet, North Carolina, for over a decade and hadn't found such an artifact. The ship's location, contents, and age of its components all matched the historic description of that ship and its ultimate demise. (Blackbeard had apparently scuttled it, with men aboard, in an attempt to reduce the number of sailors with whom he'd have to share his pirate's booty. Tim called it one of the first cases of economic downsizing.) But in a decade of work, they hadn't yet found definitive evidence of the ship's identity. For us to find and recover this artifact within four days was totally beyond anyone's expectations and almost beyond belief. I was hoping we would find the ship's bell too, but the odds of that were slim. This was just as exciting though, and having it in hand was almost enough to make the whole project worthwhile.

We hustled to get back to the dock by six so we would have time to set up for our presentation at Kodiak College. I drove home for a quick shower and picked up an empty 10-gallon aquarium and a fancy red tablecloth. By seven, we all assembled at the college, where all fifty seats were occupied. In a brief introductory statement, I outlined the importance of the *Kad'yak*: It was the only ship from the Russian American colonial period ever found, making it the oldest known shipwreck site in Alaska, and our work on it was the first professional underwater archaeological survey in Alaska. Tim gave a presentation on the ECU Maritime Studies program, emphasizing the necessity for communities to claim ownership of their maritime heritage. Then I described the circumstances of the *Kad'yak*'s sinking before turning the show over to Frank, who talked about the process of doing underwater archaeology and the hypotheses that the ECU team wished to test.

Of course we were having fun looking around the wreck, but archaeology was also about asking questions. To verify the identity of the wreck, the ECU team had a carefully designed set of sequential hypotheses to test:

A. Is it the remains of a mid-nineteenth century vessel? Or just a collection of miscellaneous junk? Does it have features typical of ships built in that era, or were any items manufactured after 1860?

B. If A is true: Is it a ship that was built in Northern Europe? Analysis of wood from the ship could identify not only the species of tree used but where the tree had grown.

C. If A *and* B are true: Is it a ship that could have belonged to the Russian-American Company? Materials found on the ship could identify it as such, even if it wasn't the *Kad'yak*.

The "*Kad'yak* artifact." Left: Steve Sellers takes a measurement to the *Kad'yak* artifact in situ. *(Tane Casserley/NOAA.)* Right: A close-up look at the artifact and the engraving. *(Stefan Quinth.)*

D. If (against all odds) A *and* B *and* C are true: Could it possibly be the *Kad'yak*? Only an item marked with the ship's name would suffice as evidence to support this conclusion.

We expected to answer some but not all of those questions. In his laconic manner, Frank talked about the mysterious object and showed a photo of it and its recovery in a bucket. In an almost offhand manner he talked about finding the ship's name written on the metal. At this point, Tim interjected, saying how unusual it was to ever find such concrete evidence—and only four days into the project.

Frank walked over to the aquarium and removed the tablecloth. "And there it is," he said casually. "You can come up and look at it if you like."

A moment of stunned silence was followed by a round of applause. People then gathered around the tank as if looking at some strange sea creature.

I kept to the background talking with friends, not wanting to upstage the artifact or the archaeologists. Some of the audience consisted of board members from the Kodiak Historical Society, the Baranov Museum, and the Maritime Museum. They were clearly thrilled. After milling about for twenty minutes or so, the dive team posed for a picture with the artifact in front of us.

"There it is," I said with a grin. "The *Kad'yak* in a tank."

CHAPTER 19

A
HARD DAY'S
WORK
UNDERWATER

FRIDAY, JULY 16, 2004: THIS morning I took my Nikonos camera with me to the *Kad'yak* site to photograph some objects in detail. Before the dive, I made a list of them on my slate, so I would not waste any time underwater. I was using a 20mm lens, which I have not previously used. Wide-angle lenses such as 15mm and 20mm have great depth of field though, so focusing them correctly is not much of an issue. While photographing, I was also going to be the "underwater model" for Stefan, who wanted to film as much of the process as possible. On this dive, I shot about half of my thirty-six-exposure roll while Stefan mostly filmed me at work.

I was particularly interested in getting detailed photos of a machinelike object we had found. Jenya and I had both looked at it but had markedly different interpretations of its size, shape, and orientation. But we both agreed that it was obviously manmade, not just a jumble of concreted ballast stones. One end of it looked like a wheel of some sort. As I photographed that portion of it, I thought it looked like a large block or sheave. It was round, with a cavity in the middle, something that looked like a support bracket at the top, and it was 4 to 6 inches wide, with a groove around the outside. I thought perhaps it was part of a block and tackle used to raise and lower ice or other equipment.

In the afternoon, Pete and I triangulated and measured a large davit to the north of the baseline, and went searching for another that I had seen and photographed last year. The latter was not to be found, and after swimming around in circles for a few minutes, we gave up. Perhaps we would find it again in the future.

I returned home that evening and found the *Kodiak Daily Mirror* in my mailbox. On the front page was a photo of the mystery object, with the name KODЯK clearly showing, under the headline "Grail confirms *Kad'yak* wreck site." The article, written by Drew Herman, was mostly complimentary.

Although I was only mentioned as having organized the discovery of the ship, I felt very proud to be associated with the project. We came a long way from my dreams and schemes of previous years to a reality I had hoped for but not ever completely expected. The work, discoveries, and fruit of future years' work were not yet a reality in my mind, but I knew many truths about it already. It would be hard, tedious work. Convincing others of the importance of supporting and funding the project would be a frustrating experience. It would be difficult, both emotionally and politically, to deal with those who denigrate and demote the work. But it would also teach us many things about our history, about ship construction and the wrecking process, and about ourselves. The ultimate result would be a dedicated museum display. Tourists, visitors, and locals would view it and learn a few things about Russian America. Perhaps they'd be fascinated by the story of Father Herman and Captain Arkhimandritov. Perhaps they wouldn't think much of it. Whatever they thought, they wouldn't know about the hard work, the years of uncertainty, the days of cold water and tired bodies, and the danger of working in such an underwater environment. They certainly wouldn't think about the dedicated researchers and all the time these people spent working on this project for little pay and no glory. But they would see the beauty of the connections between the ship, the saint, and the sailor. And that was our real legacy.

TUESDAY, JULY 20, 2004

On Sunday, the ECU group chose to test out its new ROV. None of the crew had ever used it before, so they were also joined by the two men who invented and built it. Because of the crowded conditions and since I had other pressing chores, I decided to stay ashore. Besides, if I wasn't going to be diving, I would rather not be aboard.

But they later told me that it didn't go as well as planned. In the morning, the group made a dive on the wreck site before putting the ROV in the water during their lunch break and afternoon siesta. During two short ROV dives they had trouble with the video and control link. On the third dive, a waterproof seal gave way, allowing seawater to leak into the electronic compartment and causing a flash fire. That was the end of ROV work. The builders took responsibility for the problem and carted off the components at the end of the day, with a promise to replace them. After spending close to $80,000 on that piece of equipment, Tim and Frank told me they were glad that they brought the vendors aboard for the test dive.

But yesterday I was back on board, and for the past two days Jason and I measured, drew, and triangulated various pieces of hardware on the bottom, including both cannons, a davit, and the pile of unidentifiable machinery. We spent most of a dive searching again for the missing davit I had seen last year. I had recorded its location as 20 feet from the smaller cannon, at 220 degrees, but we couldn't find any trace of it. I dragged a knife blade through the gravel, carving pie shapes over a quarter of a circle with a radius of 25 feet, with no impact of metal. We could only conclude that the shifting sand had reburied the davit.

We did, however, find another iron rod that was about 6 feet long and could have been part of a third davit. As I unearthed the end of it, I found it lying on wood about 6 inches under the gravel. This location was close to the 92-foot marker on our baseline. So far, we had only found wood from the 140-foot mark out to about the 100-foot mark. This find indicated that the hull of the ship extended further out from the ballast pile than we had previously thought, and was still buried under sand and gravel.

Back in February when Frank, Tim, and Evgenia had come to Kodiak for the exploratory dive, Frank had seen a large copper tube he later described as "like an artillery shell," which was a good 100 feet or more south of the anchors in an area I had not explored. Frank didn't make much of it at the time, especially since he was trying to find the ship, and we were diving under difficult winter conditions. This morning, he and Steve went searching for the tube again and found it 200 feet from the baseline, up on the reef. It was indeed a copper tube, 4 inches in diameter and about 2 feet long, with a heavy lead gasket at the one end and a tapered neck at the other. Tim identified it as part of a bilge pump.

"How do you know that?" I asked, incredulous.

In response, he pulled out a stack of photocopied pages from a book titled *Ships' Bilge Pumps: A History of Their Development.* I shook my head. Who would have known about such a book except archaeologists?

"It must have been a big seller," I joked.

The bilge pump was a device built to remove water that entered a ship. The narrow neck of the pump would have rested in a box in the bilge, and a piston with a leather gasket would have resided inside the tube. To operate it, deckhands would work a crank on the deck that moved the piston up and down, drawing water up through the tube and over the side of the ship. Under any lesser circumstances perhaps the pump might have enabled the ship to survive leakage, but the damage caused to the *Kad'yak's* hull when it hit the reef was so great that the pumps were overwhelmed and could not pump out water fast enough. It's amazing to me that we found the bilge pump still intact after rolling around on the seafloor for over 140 years.

WEDNESDAY, JULY 21, 2004

Today was a cleanup day. Frank had a list of various measurements and collections that he needed and parceled them out to various dive teams. Yesterday Evgenia had spent the day in the Kodiak Fisheries Research Center lab, making a detailed drawing of the artifact inscribed with the ship's name. Today, her job was to draw details of several other parts of the wood and metalwork remaining on the seafloor. I dove as her buddy and helped her with measurements.

Frank had also asked me to bring up two small artifacts we had located on previous dives. One was a spigot or faucet handle. The other, about 8 inches long, was a brass bracket of some type, and it looked like it might have held up a shelf or been used for a lamp hanger. As I waited for Jenya to finish her drawings, I turned to see

Stefan filming the ballast pile about 15 feet away. He had planned to videotape us collecting the items, but he wasn't coming over. Jenya was running low on air by then, and though I waited for a few minutes more, he still didn't come. Jenya pointed to her pressure gauge which read 1000 psi, still enough air for another minute. I indicated for her to wait, but I could detect her impatience. I wrote on my slate: *I'm okay, I'll take you to buoy.* I planned to send her up to the surface then come back to help Stefan.

No. Stay together, she wrote on hers. *What we do?*

Waiting for Stefan, I wrote on mine.

Screw it. Let's go, she wrote. For a Russian, she knew American vernacular pretty well.

Stefan still didn't show any signs of moving in our direction, so I nodded my agreement, picked up the spigot, then swam to the buoy line and picked up the bracket. Before picking them up, I briefly entertained the thought of waiting for the next dive to collect the artifacts. A cardinal rule of diving is to never put off later what you can do now. The weather could change dramatically in a few hours, and there might not be a next dive today, tomorrow, or anytime. Better do it now. I had forgotten my mesh goodie bag, so I held onto the artifacts tightly in my three-finger mitten as we ascended. I couldn't afford to drop these items; if they landed on the reef, we might not find them again.

Pete Cummiskey made another interesting find. While helping Frank measure the windlass, Pete had spotted a piece of iron protruding from the sand. Fanning away the sand and gravel with his hand, he revealed a large, rectangular iron frame, with another circular frame lying on top of it. The frame was made of 3 inches of iron, measuring thirty by 42 inches. Almost a foot of gravel covered parts of it. We speculated that it might have been a hatch coaming, the part designed to stop water from entering, although those were usually made of wood. The circular frame might have wrapped around a boom or spar at one time. Perhaps it was part of a parrel, the apparatus that attaches the spars to the mast. Or maybe the opening to the chain locker.

In fact, this sort of guesswork was becoming more and more the norm. As we neared the end of this project, much of what we saw couldn't be positively identified. It might be this, it might be that, or it might be something else. East of the ballast pile was a jumbled area of rocks and mangled metal, most of which were totally unidentifiable. We measured, drew, and triangulated various portions of it, but without knowing what they were, the effort seemed rather futile. I was ready for something bigger to chew on.

Last year while searching for the mysterious cannon I had seen a large metal spar or rod, about 20 feet long, at a site several hundred feet away from the main ballast pile, in an adjacent channel to the east. At that location, I had picked up a solid brass rod about 6 inches long. It was similar to the drift pins in construction, but much heavier and thicker. Despite the distance from the main wreckage, and the lack of other significant artifacts nearby, I believed these items must have come from the *Kad'yak*. Frank had put off searching in that area as long as he felt we had much work to do at the main wreck site, but now he felt we were ready to investigate it.

This morning while I had worked with Jenya, Frank sent three teams of divers scouting in that direction. One team stretched out a 300-foot measuring tape due east into the adjacent channel until it ran out without finding anything of significance. The next team went directly to the site I had described, on the north edge of the channel up against the reef. Upon returning they excitedly described their find. There was not just one spar but four or five, plus lots of chain and other metal machinery. The best part was that they had found evidence of the ship's rudder. Several sets of pintles and gudgeons, the upper and lower parts of the rudder hinges, were lying about close together. Shipworms had long since destroyed the rudder, but the metal clearly outlined where it had landed. This was another major breakthrough.

During our discovery dives in July last year, Steve Lloyd had brought up a heavy brass strap with a drift pin embedded in it. After scouring the contemporary whaling ship *Charles W. Morgan* during a trip to Mystic Seaport in Connecticut, I had concluded that the only similar part to the *Kad'yak* artifact on board was the pintle, attaching the rudder to the ship. We still didn't know exactly where Steve collected that part, as he didn't record the information, but I had assumed it was at the ballast pile. Without that knowledge,

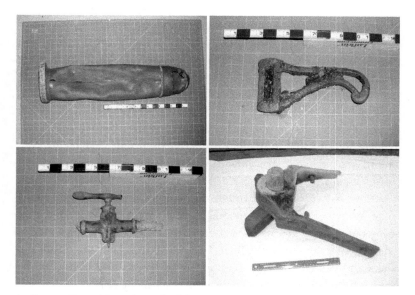

Artifacts collected from the *Kad'yak* for preservation. Top, left: The bilge pump. Top, right: Unidentified bracket. Bottom, left: A spigot, possibly from a water barrel. Bottom, right: Pintle (left) inserted into the gudgeon (right). Only the gudgeon shows remnants of the straps by which it was attached to the vessel. Squares in background mats are 1 inch. *(Bottom, right: Frank Cantelas.)*

we couldn't identify it. The location of the rudder parts suggested several possibilities. One was that the ship could not have stretched out over such a large distance; therefore, it must have initially come to rest on the top of the reef and then broken into pieces over time, and fallen off the reef in several directions. A second conclusion was that metal machinery near the rudder could be parts of the steering mechanism.

Mark Blakeslee was also on board today with his Phantom ROV, which he had named *Rosebud*, a humorous reference to the sled in *Citizen Kane*. It was much larger and more robust than the ROV that Tim had tested a few days earlier. In the afternoon, Mark put *Rosebud* into the water and flew it over to the new site. Mark and I had searched for the *Kad'yak* with *Rosebud* in 2001 and 2002, but we had been looking farther to the north and west in the bay. The closest point we had searched was on the far side of the reef just north of the ballast pile. If we had flown up over the reef and down into the next basin, we would have hit it. Would we have seen it? Recognized it for what it was? At the time we were searching for something more obvious. Underwater in the lights of the camera, rock, metal, and wood all looked the same, covered with ubiquitous pink encrusting algae, half buried in sand and gravel. Was that a rock or metal? With my diver's eyes, I could distinguish the two by looking for the telltale color of rust, the copper-yellow color of brass and bronze, or the brick-red-green of copper, or by discerning the three-dimensional aspects of items that didn't look quite like their natural background. The ROV's video camera, however, provided only two-dimensional tunnel vision.

Now, as we viewed through the ROV's camera what we knew to be wreckage, the only obvious features were straight lines that stood out slightly against a backdrop matrix of fractal chaos. Upon closer inspection, straight lines became chains encrusted into the rocks and square kelp fronds became gudgeons and pintles while unusually symmetrical rocks became machinery parts.

It was now obvious, with only three days of ship time left, that we needed to move our operation 100 yards east and investigate this new area, which we thought was the stern section. That would be the plan for tomorrow. I had another pressing matter to deal with.

I looked down at my hand where a large boil had formed, yellow in the middle with blue-black edges. A month ago, I had been collecting hairy crabs from the tidepools for a project. I usually wore a heavy rubber glove on my right hand to turn the barnacled rocks over, leaving my left hand free to pick up the little crabs. That day the sun was out, and I was concerned that the crabs in my bucket would get too warm, so I set the bucket down at the water's edge. A few minutes later I looked up to see my bucket drifting away. At that moment I forgot one of the rules of tidepooling: Don't run! Be careful! Kelp is slippery! On the third step my foot slipped out from under me and I fell backwards, landing my ungloved hand on top of a group of sharp-edged barnacles. Leaving the bucket on a rock, I clambered back across the rocks to my car where I had a small first aid kit and swabbed the bleeding cuts with alcohol and Betadine. I had debated going to a doctor, but it was Saturday that day, which meant the Emergency Room was my only choice, and I knew that would be expensive. I decided to wait and see.

The cuts seemed to heal over the next few days, so I just kept replacing the bandages. After two weeks it was still tender but seemed to be healing over. Then on the first day of diving on the *Kad'yak* this year, the wound broke open. As I was pulling the tight drysuit cuff over my hand, I had felt a sharp prick. I peeled off my bandage and saw something white sticking out. With the forceps from my Swiss army knife (another use!) I removed from my hand a piece of barnacle shell about the size of a pencil eraser. *After years of studying crustaceans*, I thought, *I am finally turning into one.* The gang made jokes about me growing claws. After a few days, the wound healed over again, and I stopped wearing a bandage for a week. Then it got ugly.

So today I finally decided to take care of it once and for all. From the *Big Valley* in Icon Bay, I called the doctor's office on my cell phone. They agreed to stay open until I got there at night. I didn't want to wait until the next day and possibly miss a day of diving. That evening Dr. John Koller at the Kodiak Ambulatory Clinic opened the wound and removed another piece of barnacle and some sand, as well as surrounding tissue that was becoming necrotic. I should have done this weeks ago. Sometimes I am just too stubborn for my own good.

THURSDAY, JULY 22, 2004

I went down to the *Big Valley* for breakfast as usual but decided to take the day off. In addition to a sore hand, I had come down with a scratchy sore throat. Almost everybody on the boat had gotten sick, and most fingers were pointing at Tim as the culprit who had brought it aboard. I hated to miss a day of diving, but my body was crying out for rest. Besides, it was blowing 20 knots from the northeast, and the forecast was for 10-foot seas. Because of its southeast opening, Icon Bay would be partly sheltered from that, but the swell had been already 3 to 4 feet there yesterday afternoon, making it difficult to climb up on the stern ramp. Frank had been washed off three times. I had doubts that they would be able to dive today, so it was probably a good day to rest. Sure enough, I heard later that when the *Big Valley* had arrived in Icon Bay, the waves were too large for safe entry and exit from the boat, so the crew returned to Kodiak without diving.

Instead of diving, I immersed myself in our good week of publicity. Over the past two days Tim had been on the phone with NOAA public affairs, the Office of Exploration, and science reporters at the *New York Times* and *San Francisco Chronicle*. We had also done lengthy interviews with our local reporter for National Public Radio. Our goal was to craft a feature article about the *Kad'yak* and even include the history of Saint Herman and the ship's sinking. And for the cherry on top, our website was also completed this week, hosted by Alaska Department of Natural Resources.[11] So though I couldn't be out at sea, I still stayed busy, the *Kad'yak* always on my mind.

FRIDAY, JULY 23, 2004

Today was our last day of diving at the *Kad'yak* site. I had hoped to finish photographing some of the artifacts, but Frank wanted me to help Jenya with some measurements of the newly found artifacts at the stern site. So during our first dive today, we triangulated, measured, and drew three long metal spars or beams. Then began the saga of the slate. At one point, Jenya's mechanical pencil stopped working, or maybe it ran out of lead. I tried to give her mine but could not detach it from my slate, so I unclipped the whole slate and handed it to her. Unlike her slate, mine was buoyant, and when she accidentally released it a few minutes later, it floated away. Somehow, I managed to make her pencil function again, and we finished our work.

Up at the surface, there was still a 4-foot swell from the previous day's storm, which made getting back onto the *Big Valley* a challenge. I floated just behind the boat, waiting for the right wave and kicking backwards to keep from being drawn under the boat, as the stern rose up and down in front of me. Finally, a large wave came along as the dive platform at the stern of the *Big Valley* sank down in front of me. As the wave lifted me up, I stepped easily onto the dive platform and held on as the stern rose up in the air with me hanging on to it. *Yes! That's how it's done*, I thought, images of rodeo cowboys in my mind. No sooner had I done that than some killjoy on deck pointed out that my slate had just floated by behind me. After that easy boarding I had second thoughts about getting back into the water. *If I didn't know better, I'd think that you had put that there just to make me repeat that performance*, I thought. But I needed my slate. So as the stern dropped down again, I flopped back in and retrieved it. It took me several tries to get back onto the dive platform again without getting washed off in the swell.

During our second dive of the day, I carried my camera and took some photos as we finished our measurements. As I swam around, I discovered a round greenish-black ball, about 4 inches in diameter, lying on the reef. *It can't be!* I thought. *Is that... a cannonball?* I couldn't believe my luck. It must have rolled off the ship as it sank. I had been hoping for a while that I might find one of these. Carefully, I took two or three photographs, then I reached out with one hand to clear the ball of some debris. Instantly, it imploded into a cloud of dust, swirling and mixing in with the other particles in the water. I was so shocked, I couldn't move. There was no way I could've destroyed an artifact just like that. Then I realized—the ball was not a cannonball at all but just a spherical piece of algae. The only damage it could do was to blow up my wishful thinking.

With only a few minutes of air left, I swam to the stern section to see what else might still be there. Beyond the spars, a 20-foot length of chain draped down the side of the reef, ending in a large concretion, probably consisting of more chain.[12] Aside from numerous drift pins of various lengths, I only saw a few oddly shaped pieces of brass. There were likely many more small items we had not had time to inspect carefully, though I was sure that this was exactly where I had seen the "spar" in 2003 and had picked up a short but heavy chunk of brass. I hadn't seen all the chain links during our first visit last year because I didn't really know what we were looking for or what it would look like. Now, after working the shipwreck site for the last two weeks, I had a much keener eye for artifacts.

Back on deck, we were all tired but happy to have finished the project. That evening, Gary treated us all to another adventure. Instead of heading back to Kodiak, he turned the *Big Valley* up the channel toward Ouzinkie and across Shelikof Straight to Katmai National Park on the Alaska Peninsula. Passing through Whale Pass we saw a pod of fin whales, and a humpback whale breached six times in as many minutes while we all whooped and hollered with excitement. Dall porpoise jumped and surfed in our bow wave as we sped along. Otters eyed us cautiously from the safety of kelp beds as we cruised by.

As the late summer sun set behind the mountains, and the stars came out above, we drank wine on the back deck and watched the phosphorescence in our wake. Then we all turned in and slept on board. The gentle rocking of the boat lulled me into a deep sleep as the *Big Valley* plowed through the seas all night.

SATURDAY, JULY 24, 2004

When I woke this morning, I felt totally refreshed. I was thrilled that we had finished our survey and happy to be aboard ship with people I had shared an incredible adventure with and come to know like best friends. For the next few days, we would enjoy a brief Alaskan vacation to celebrate our success.

Early in the morning we cruised into the incredible steep walled fjord of Geographic Harbor, ringed by precipitous 4,000-foot cliffs. Waterfalls fell from the heights into mist-veiled valleys. Before us, freshwater creeks flowed into the ocean, forming a wide delta. We climbed into two skiffs and motored into the beach, where we watched bears chasing fish in the tidal creeks. The bears didn't mind us as long as we stayed in a group and kept our distance. Then in the afternoon we traveled to outlying islands, where we inspected an ancient Aleut midden, with mammal bones and fish parts protruding from the hillside. We were careful not to disturb it, though. It may look like somebody's ancient refuse pile, but some future archaeologist might find important artifacts there, illuminating the seven-thousand-year history of Native occupation.

While some of us explored the middens, Bryce, Dave, and Steve went fishing and caught three halibut. That evening, the gray skies parted for a few hours, long enough to have a celebratory party on the back deck of the *Big Valley*. Gary heated up a charcoal grill and barbecued the halibut for dinner. Each of us took turns telling stories, singing songs, and reciting or reading poems. I sang an acapella version of a song I had written and recorded with Waterbound, one of my Kodiak musical groups:

> *Watching the wildlife in Kodiak, AK;*
> *Watching the wildlife in Kodiak, AK;*

You better be careful and not turn your back,
When you're watching the wildlife in Kodiak, AK.

I couldn't stop smiling as I listened to the voices around me. Though the evening grew cold, I had a warm feeling inside. I was home, and these were my colleagues, my friends, my family.

THE NEXT DAY FOUND MOST of us hunkered down in our bunks as the *Big Valley* beat its way back to Kodiak through 30-knot winds and 8-foot seas. In late afternoon, we cruised back into the calm waters of Kodiak Harbor. Our excitement had slowly abated over the next two days as we unloaded the *Big Valley*. We packed up scuba gear, cameras, and personal belongings, and prepared artifacts for shipping. Poor weather caused several flight cancellations, but one by one the team members said their goodbyes and returned home.

The ending of such projects usually made me sad, knowing that I might not see or do anything similar for a long time, if ever. But at the same I felt a great sense of satisfaction and accomplishment. We achieved so much more than we had initially expected. We had found many large artifacts not seen in 2003, including one verifying the identity of the ship (see Appendix C). Some of the artifacts were intact and in good shape for initial preservation. And we had seen and documented large expanses of timbers and found that they extended further under the sand than we could excavate this year. We had learned that the site was very dynamic—it may be covered and uncovered by shifting sand and gravel at any time. We had also found the stern section, or at least parts of it, separate from the rest of the ship. This would provide many clues about how the ship foundered and eventually disintegrated.

Aside from these concrete objectives, we had achieved other equally important though less tangible results. We had established scientific observation and historic documentation as priority "uses" of shipwrecks in Alaska. We had set the tone and defined the standard for all future marine archaeology in Alaska. We had included and engaged the public and shared with them the thrill and excitement of our work. We had established good personal and professional relations with community leaders, who would be advocates for responsible stewardship of cultural resources. Best of all, our dive team had grown to know one another personally, to trust each other as dive buddies, and to respect all as professionals. We were more than colleagues now; we were friends. And we'd had a hell of an adventure.

CHAPTER 20

FOR
WHAT
IT'S
WORTH

THE ULTIMATE QUESTION ABOUT THE *Kad'yak* is: Why does it matter? After all our work, what is the value of the knowledge gained about the ship's construction, its history, and its demise? What is the purpose of finding it, surveying it, and documenting its condition? Was it a worthwhile endeavor? And who cares? Although that may sound a bit dramatic, these questions are commonly asked not only by the public but by legislators and grant reviewers who make decisions about spending scarce resources. While the answers seem obvious to me, they may not be to everybody, so it is important to explain them.

What is the significance of a single shipwreck such as the Kad'yak?

The *Kad'yak* was typical of ships that sailed in Alaska in the 1850s, and the wreck itself was a commonplace occurrence. The fact that it was documented, found, and surveyed makes it unique among the pantheon of Alaskan shipwrecks. It has much to teach us about the history of seafaring in Alaska.

Alaska is essentially, and historically, a maritime nation. Many Alaska Native communities were located on the seacoast and along rivers, and they depended greatly on water for hunting, traveling, and communication. Native cultures developed highly seaworthy crafts including kayaks, baidarkas, and umiaks, which are large, open boats made of marine mammal skins. The history of settlement in Alaska by Western non-Native cultures is a history of seafaring men and nations. Ships were the only means of travel and transport to Alaska for centuries. The first Russian explorers came by ship and met Alaska Natives in their own skin boats. Many later settlers came to fish and built large coastal communities to support this activity, including Petersburg, Juneau, Kodiak, and Dutch Harbor. Even today, much Alaskan commerce is conducted by ships, which are more efficient for carrying cargo and still carry most of Alaska's oil to refineries in other states. Thus, shipwrecks were an inevitable and unavoidable part of Alaskan History. Between 1740 and 1868, over 170 Russian vessels sailed in Alaskan waters, with over 30 percent of those ships reported lost at sea. And that number does not include the American and British ships that sailed in those same waters.

The frequency of shipwrecks was the result of a number of factors. First and foremost of those factors is Alaska's geography. The coastline consists of a series of islands, inlets, and fjords, liberally sprinkled with unseen underwater reefs, rocks, pinnacles, and shallows. Early mariners had to be either extremely cautious, or extremely lucky, to navigate in and out of Alaskan ports without incident. Even highly experienced navigators such as Captain Arkhimandritov fell victim to uncharted and unseen dangers. Second, the lack of adequate maps to guide them made such journeys even more treacherous. Prior to publication of Tebenkov's *Atlas*, navigators had to rely on their own knowledge and information passed to them from previous voyagers, much of which was erroneous. Even with good maps, ship captains sometimes made errors, running afoul of well-known and well-charted hazards, as in the case of the *Exxon Valdez*. Third, of course, is the weather. The North Pacific Ocean and Bering Sea are breeding grounds for major storms that sweep ashore with extreme force, commonly bringing winds exceeding 50 knots. Winter weather in Kodiak consists of a series of storms with winds from the southeast, generating huge swells and waves that crash into bays and fjords. Icon Bay, facing southeast, is particularly exposed to such storms, and thus it is even more remarkable that parts of the *Kad'yak* remained as intact as they did. Finally, many of the earlier Russian ships were poorly constructed and manned. Crews often consisted of men who had never sailed before and didn't know about maps or compasses, and ships constructed in Okhotsk were so crudely built that they were almost inoperable. The *Kad'yak*, though, was a much newer ship and built in Germany to high standards, so its fate was not due to faulty construction.

What have we learned by surveying and documenting the condition of the wreck site?

The artifacts that we found and collected tell us much about the wrecking process. After drifting for three days, the *Kad'yak* came to rest in Icon Bay. It is remarkable that it traveled as far into the bay as it did. In order to arrive at its final location, the ship had to pass over several outer reefs whose tops come up to within 45 feet of the surface. Most likely, as the ship drifted to the north, its cargo of ice continued to melt, allowing it to sink lower in the water. Finally, it would've grounded out on a reef top. There, its top deck would have been submerged under about 20 feet of water, and over time, as a result of continuous pounding by waves and storms, the ship was probably pushed to the northeast along the top of the reef, breaking off the rudder and stern section and dragging the steering mechanism and chains along the reef. Eventually, it must have pushed off the reef top into the channel bottom where it remained and broke apart under attack from weather and water. With storm surge and currents moving the sand and gravel, items on the seafloor would have been tossed around as if in a washing machine. The masts may have broken off and drifted away. Any organic materials, including most of the wood, would be destroyed by shipworms and bacteria. Metallic objects lasted longer; those made of iron rusted and were concreted into unidentifiable shapes. Items made of copper alloys fared much better and remained largely intact. Heavy items lay where they fell; smaller and lighter items were pushed around by waves and currents.

One of the most important things we have learned from the *Kad'yak* wreck site is that shipwrecks can survive for long periods of time, even under the extreme conditions that occur in Alaskan waters. Had it sunk into anoxic muds of a calm, protected bay, it would have been much better preserved. Nonetheless, the *Kad'yak* has lain on the bottom for over 150 years, yet many items are still present and identifiable. In order for archaeologists to extract information from the wreck site, it is important that artifacts remain in their original locations, undisturbed by humans, until they can be documented properly.

What were the benefits of exploring the Kad'yak *wreck site?*

The *Kad'yak* expedition of 2004 was the first professional underwater archeological survey ever conducted in Alaska. As such, it set the standard and a precedent for all future marine archaeological surveys in the state. Many invaluable lessons were learned. Conflicts were inevitable, and the ensuing arguments, both public and private, over credit for the discovery, ownership of the wreck site, and disposition of the artifacts illustrate the kinds of obstacles that will be faced in future such efforts. We can only hope that professional archaeologists will learn to predict and avoid many of these conflicts. We also hope that the public, through exposure to news articles about these discoveries, will be made more aware of the value of protecting historical sites. Although the *Kad'yak* only sailed in Alaskan waters for eight years, it participated in many historic activities, including an around-the-world voyage, trade missions to California and Hawaii, and the important and lucrative ice trade that was unique to Kodiak.

An important aspect of the grant-funded project was public outreach. The presence of "outsiders" doing research on cultural artifacts in a small community generates suspicion as well as interest. During 2003 and 2004, we conducted several public meetings in both Kodiak and Ouzinkie. These meetings served to share our findings, increase awareness of the project, and communicate the significance of the *Kad'yak* to local history, especially its association with Saint Herman. Such meetings also helped to show that the scientists and divers were serious professionals, whose primary concern is the preservation of the wreck site, not removal of artifacts. All of this contributes in developing a sense of "ownership" among the local community and ensuring that the *Kad'yak* wreck site is preserved for posterity.

In addition to public meetings and press releases, the project included funding for collaboration with two local teachers, Balika Haakanson and John Adams. Using materials provided by the scientists, the teachers developed lesson plans for students in the Kodiak Island Borough School District, focusing on the history and science of the *Kad'yak*. Introducing these topics to children is the best way to ensure a future wherein Alaskan history is well known and the importance of historical resources is recognized.

CHAPTER 21

THE
LOSS
OF THE
BIG VALLEY

SIX MONTHS AFTER OUR HISTORIC expedition, on January 15, 2005, Gary's *Big Valley* had set out from Dutch Harbor for the opening day of snow crab season. For over twenty-four hours it pounded to the northeast through 35-knot winds. The storm blew steadily from the south, creating 15 to 20 foot waves, so that the ship rolled from side-to-side "in the trough" most of the previous day and all night. It was uncomfortable but not necessarily dangerous. On board were thirty-five of the big 700-pound crab pots, and 11,000 pounds of bait. Another twenty pots were being carried by another boat, although it was against the law for them to do so. The *Big Valley*'s legal limit of pots was fifty-five, but with this seasons' crab quota set at only 25 million pounds, Gary knew it would be a short season, and he might not get the chance to go back to Dutch Harbor for the rest of them. It was almost 500 miles for a round trip, and would require at least two days, under the best of conditions.

On board the ship were four crewmembers: Josias Luna, resident of Kodiak and father of three, had been a crewman on the *Big Valley* for many years; Danny Vermeersch, a Dutch citizen, had been fishing with Gary on and off over the past few seasons; wiry Cache Seel was over 6 feet tall and made of mostly solid muscle; and Carlos Rivera, the youngest of the crew, was there for his first snow crab season on the *Big Valley*. In addition to these crewmembers, they carried a guest. Aaron Mars was a young filmmaker and photographer from Louisville, Kentucky. Aaron had bummed with a friend into Kodiak the previous summer and had fallen in love with the place. One trip on a fishing boat, and he was hooked; he had to come back and film a documentary about crab fishing. When Gary offered him a berth, Aaron loaded his cameras on board and never looked back.

At four in the morning, the boat was lugubriously slogging through the seas 70 miles west of the Pribilof Islands. Gary and most of the crew were asleep in their bunks while Carlos was on the wheel watch. Sitting in the dark, surrounded by fathometers, radios, radar and other electronics, could be a mesmerizing experience. Thank goodness for the bridge alarm, which buzzed at five-minute intervals just to remind him to stay awake.

Cache woke up suddenly, in a strange position. He had been sleeping in the stateroom next to the galley, one bunk below Aaron. Now, in the dark, he was standing up. He knew instantly what had happened. A wave had caught the boat and rolled it over 90 degrees onto its side. Instinctively, he crouched down and grabbed the survival suit that always lay at his feet. Struggling against the sickening bob of the ship, he managed to slip into it and zip it up. He stepped out of his bunk into a chaotic world turned on its side.

Where was the door? It took only a moment to realize it was no longer beside his bunk but now over his head. He reached up for it, braced his feet against the bunk and pulled his body up through the door into the passageway. Below him, Aaron had woken and was yelling for help because he couldn't reach the doorway. Cache reached down, grabbed Aaron's arm, and hoisted him up out of the stateroom. It was dark inside the ship, and they were disoriented, but they could hear others yelling. Outside, the crab lights were still on, creating a dim glow that highlighted the shape of the galley windows above them. They worked their way back toward the doorway, almost falling into the aft stateroom in the process.

Gary had also been thrown out of his bunk on the upper deck behind the bridge. Somehow, he made his way down into the bridge, then up to the ladder, and sideways into the galley. From there, he crawled into the pantry where the remaining survival suits were stowed and began throwing them back down the alleyway.

Getting into a survival suit is no easy trick. It's awkward, tight, and clumsy to zip up. And once inside the survival suit, you have extremely limited mobility. Its buoyancy causes you to float on top of the water. Swimming is practically impossible because you can't push your arms down into the water or kick. About the only thing you can do is turn over on your back and paddle with your arms, as if you were making snow angels. Donning a survival suit in the dark, inside a boat that is lying on its side in a rolling sea, is extremely difficult. And doing it in the water is impossible. As a practical rule, if you don't have your survival suit on and fully zipped up, when you go into the water, you will die. Period.

Even after exiting from the doorway, the men couldn't climb up over the house. With the ship lying on its starboard side, the weatherbreak, a 10-foot-tall bulkhead on the port side, was over their heads. The only way out was to go down into the water. Cache slid in, working his way back toward the stern. Up on top, someone was struggling to release one of the rafts, but the canister wouldn't open. When it finally popped open, the wind blew it into the rigging where it tangled hopelessly. Within minutes, the ship rolled completely over, its hull rising up through the waves like a drunken whale.

Cache couldn't see Gary but heard his voice shouting through the howl.

"I've got the EPIRB, it's working. The Coast Guard will be here. Hang onto the boat!" At this moment of desperation, Gary was probably thinking of the crew, trying to encourage them to stay together.

The pounding waves pushed Cache back toward the rudder. He could no longer hear nor see anyone else. For as long as he could, he hung on to the rudder brace. As the ship settled lower in the water, waves washed over his head. Each time the ship bobbed, it dragged him underwater, head down, feet buoyed up by the air in his suit. Each time he came up, he spat out seawater and gasped for breath. In the freezing water, he fought to remain conscious.

After what seemed like hours, Cache saw a light. At first he thought it was a boat but then realized it was the strobe light on the raft. If he could get to the raft, he might have a chance. It was his only hope. Releasing his grip, he swam as hard as he could toward the light. The wind and waves pushed him backwards. From somewhere deep inside himself, he pulled out the strength to swim. An hour later, he finally reached the raft and managed to pull himself in. As the waves threw him about, he lay in the bottom of the raft and vomited seawater.

THE BIG VALLEY'S EPIRB SIGNAL was detected by Coast Guard Headquarters in Juneau. But the signal was weak and they could not pinpoint its location. Officers called the Kodiak Air Station and alerted them that a ship was in trouble. But rather than launching a plane, they were asked to wait for confirmation. An hour passed before Juneau called back, confirming the Big Valley's position. Soon a C-130 search aircraft roared down the runway from Kodiak Air Station and turned westward. The Big Valley was 500 miles to the west, over three hours away, across the Alaska Peninsula and out in the Bering Sea.

At about 9 am, Cache heard the helicopter. As it hovered over him, a rescue swimmer jumped out. Cache rolled out of the raft and climbed into the rescue basket, which lifted him up out of the water. The helicopter spent a few minutes looking for other crewmen but did not find any, and soon it headed back to Dutch Harbor. Later in the morning, another fishing boat arrived on the scene and began searching for bodies. Only two were found.

By evening, everyone in Kodiak had heard about the accident. Friends of Gary and the crew gathered to wait for news. It wasn't until the next day that they learned that the bodies of Josias and Carlos were recovered. Danny, Aaron, and Gary were still missing. Nobody but Cache Seel had managed to get into a survival suit. It was devastating news.

The following weekend, a wake was held at Gary's house. Gary's best friend and fishing partner Ian Bruce was so distraught he couldn't talk. He and Mark Blakeslee had both gone to the Bering Sea with Gary to fish for king crabs in October, but both were fortunate not to have been aboard this time.

WEEKS LATER AN OFFICIAL INQUIRY was held by the Coast Guard. The previous year Gary had added the salon room to the top of the upper deck, making the ship not only heavier but top heavy. As required by law, he had a new stability check done on the boat and a letter of compliance prepared by a naval surveyor. The letter stated that his weight limits were twenty pots and 7,000 pounds of bait, much less than the thirty-five pots and 11,000 pounds he was carrying. According to the Coast Guard, the *Big Valley* was simply overloaded. It was an accident waiting to happen. The lawyers and insurance companies would write it off to greed—another example of a fisherman crossing the lines to make more money.

Those familiar with the boat had other opinions, however. During the October crab season, Mark Blakeslee had been on wheel watch late one night. Next to the skipper's chair, a TV screen showed scenes taken by cameras in the engine room and on the back deck. Looking at the screen, Mark saw that there was no water coming out of the opening in the crab tank hatch. Immediately, he woke Gary, who went down into the engine room to fix it. After a few minutes, the pump came on, and water began pouring out of the hatch once more.

When a ship's fish hold is empty, the ship floats high in the water and bounces around like a cork. Heading out to fish, most skippers "tank down" the hold by filling it with water. This lowers the boats center of gravity, giving it a smoother, more comfortable ride. Unfortunately it also gives it less freeboard and makes it less buoyant. As the ship rolls from side to side, water in the hold sloshes back and forth; if there is an air space in the tank, the water becomes a "free surface," and its movement shifts weight from one side of the boat to the other, destabilizing the boat. The only way to prevent that is to keep the hold full by running the pump constantly so that the tank overflows continuously.

After the *Big Valley* sank, Mark speculated that the pump had probably quit working sometime during the night. Carlos, the green crewman, might not have noticed it. After a little while, enough water would have sloshed out of the tank to create a free surface. It's possible that when a wave hit the *Big Valley*, leaning it over, the water all moved to one side, creating momentum that carried the ship over too far for it to recover.

THE LOSS OF THE BIG Valley and most of her crew was a terrible tragedy, not only for the families of those lost but for the community of Kodiak, and the scientific community as well. Over the last decade, other scientists and I had chartered the Big Valley for many research cruises. Each of us came to know Gary Edwards as a friend and kindred spirit. The Big Valley wasn't the newest or cleanest boat in the fleet, but Gary made it the best boat to work on. Gary made everybody feel supported and comfortable. He made cappuccinos for us on board each morning. If we were too busy to eat, he brought our food out to us on the deck. He rigged up a hot-water shower on deck for us to use after diving. He built specialized equipment such as an A-frame on the stern for launching my camera sleds. Gary was a real Renaissance scholar who loved great art, literature, and music. He hung interesting artwork all over the boat and kept a shelf full of great books at hand. He went out of his way to ensure that we had the best possible experience on his boat. Gary touched each of us that knew him in many different ways, and we are all deeply saddened by his loss. But in addition to his friends and family, he also leaves behind a generation of scientists and graduate students who worked with him, learned from him, and came to love the sea as he did. He facilitated our explorations of the unknown. The oceans, and those who study them, have lost a great friend.

The galley of the Big Valley was adorned with an eclectic collection of artwork. Sepia-toned images of square-rigged ships, African Natives, Southeast Asian islanders, and various great explorers hung haphazardly from the walls. I asked Gary once what was the theme of his collection. He thought pensively for a moment, then replied, "Voyages of discovery."

Speaking for the scientists, students, and others who sailed and worked aboard the Big Valley, our journeys with Gary were all truly voyages of discovery.

We will miss him greatly.

AFTER LEAVING KODIAK, BRYCE KIDD took up welding and metal sculpture. A year later, he returned to Kodiak with a large memorial statue for the Big Valley and its crew, showing an anchor intertwined with kelp. Graciously, the City of Kodiak granted permission and land for it to be installed near the Kodiak waterfront. The statue also bears a plaque, listing the names of the crew, and a prayer:

In honor of the crew of the F/V *Big Valley*,
Lost, January 15, 2005:
Captain Gary Edwards
Josias Luna
Danny Vermeersch
Aaron Marrs
Carlos Rivera
And Survivor, Cache Seel

In their Memory,
May There Only Be Beautiful Things

On a drizzly afternoon in August, about two dozen people assembled at the waterfront for a brief ceremony to celebrate the sculpture, remember our friends, and drink a final toast to the *Big Valley* and crew. Eight months later, NOAA created a weather buoy in their honor named, of course, *Big Valley*. It was deployed out in the Bering Sea, west of the Pribilof Islands, in order to gather weather and sea conditions in that area and make them available to the fishing fleet. Maybe it will help to prevent another tragedy.

The 2005 crab season was the last open-access crab season in the Bering Sea. The following year a new management system called "crab rationalization" was put into place, which assigned quotas to boats based on their previous years landings, and allowed fishermen to sell, lease, or pool their quotas. No longer were they required to race each other to the grounds, fish as fast as possible in terrible weather, and finish the season in four or five days. Instead, they could wait out the weather and fish any time within a three-month winter season. Over 250 boats and 1,200 fishermen participated in the crab season of 2005, but in 2006, only 65 boats and less than 350 fishermen participated. The cowboy years were over; the range had been fenced. The *Big Valley* was the last boat to sink in the open-access Bering Sea crab fishery. A year later, it would not have happened.

EPILOGUE

OVER THE PAST TEN YEARS, I have told the story of the *Kad'yak* to anyone who would listen, until it began to seem like ancient history to me. During the intervening decade, many changes have occurred. After living and working in Alaska for twenty-two years, I left Kodiak, and my job with NOAA, in September 2006 to become an Associate Professor at the University of Massachusetts, Dartmouth, School of Marine Science and Technology. Although I missed Kodiak greatly, I looked forward to a new chapter in my life as a university professor. In 2009, after only three years in Massachusetts, I moved to the University of Maryland Eastern Shore (UMES), where I am now a tenured Full Professor of Marine Science.

In 2015 I started an AAUS diving program at UMES and have been diving regularly with my students since then. Some of our dive sites are shipwrecks over a hundred years old off the Maryland coast. We dive there to study how fish use the shipwrecks as habitat. Every time I dive on those wrecks, it reminds me of my days investigating the *Kad'yak*. Looking at the decomposing wood and metal artifacts, I would build a mental picture of what the ship may have looked like. Having learned from the archaeologists, I'd draw pictures of the wrecks on my slate, with notations about compass directions and distances, that would help me understand the layout of the wreck and orient me when I reach the bottom. It amused my students at first, but they quickly learned the value of having a map of their dive sites. I have had many adventures in my life, but I will probably never again have an adventure that is as exciting or as much of a roller-coaster ride, physically and emotionally, as the search for the *Kad'yak*. While writing this account, I had to reread my notes, the historical documents, and all the related news articles, which was overwhelming and brought back feelings and emotions I haven't felt for years: joy, mystery, anticipation, frustration, anger, resignation, excitement, elation, and the final satisfaction of success.

Despite my best efforts, we were never able to obtain additional funding to recover and preserve additional artifacts from the *Kad'yak*. It is unlikely that there will be any further exploration or investigation of the wreck site. Any such plan must take into account the costs—including materials and labor—of recovering and preserving the multitude of artifacts that will be discovered. Our initial estimate of the cost for a Phase II *Kad'yak* expedition was $1.5 million, and that was extremely conservative.

I am a biologist, not an historian. I took the same approach to the *Kad'yak* that I take to all of my scientific investigations: Find an interesting problem, break it down to a series of simple questions, and try to solve them one by one. Science and history are both detective processes—we find pieces of a puzzle and try to put them together in a way that makes the most sense. Eventually, if we are lucky, we start to see a bigger picture. We may never see the whole picture, but if we can see enough of it, our minds fill in the missing pieces with potential solutions. I suppose that is why I became a scientist, because my work is never done. Even if I can answer one question, many more appear that may lead off in any number of different directions that I had not previously imagined.

Even the *Kad'yak* isn't finished, although my part of it may be. Perhaps other divers will now go there, use my maps, and enjoy seeing what I saw. As they swim along the bottom, they'll see pieces of the puzzle—a pile of ballast stones, an anchor here, a cannon there, a davit poking out of the sand—and in their minds they will see the outlines of a ship, emerging from the gravel and sand, reforming itself whole in front of them as if out of the mists of time.

The *Kad'yak* was built for use in Kodiak, named for Kodiak, sailed out of Kodiak, sank in Kodiak, and became one with the island for which it was named. As we raise its remains, we also raise our awareness of its place in Alaska, from mystery to history. When I dream at night, I see the *Kad'yak* sailing down the channel, its sails unfurled, the waves lapping at its hull. In my mind's eye, the *Kad'yak* still sails, through the times, places, and events in the lives of real people who once walked the streets of Kodiak. It brings history to life.

SUPPLEMENTARY
MATERIALS

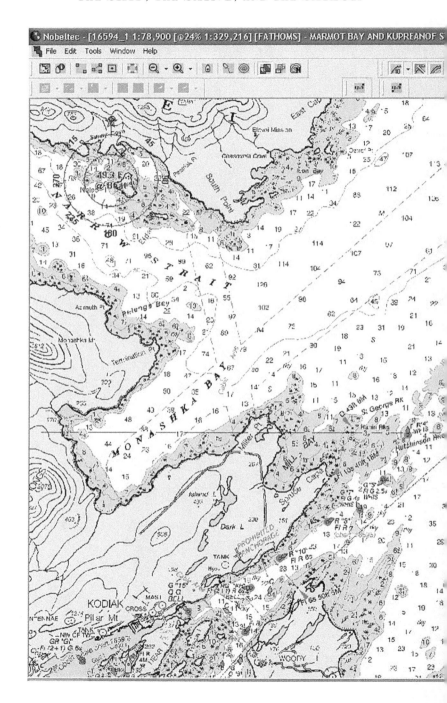

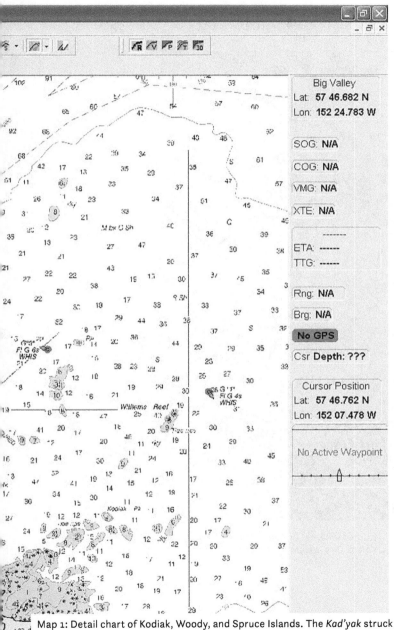

Map 1: Detail chart of Kodiak, Woody, and Spruce Islands. The *Kad'yak* struck bottom at Kodiak Rock, northeast of Long Island, and ran aground in Icon Bay, at the southeast end of Spruce Island.

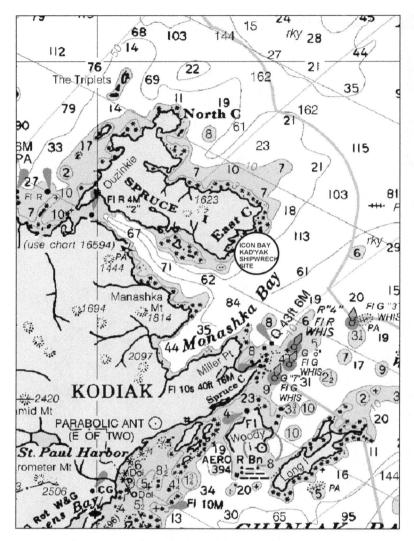

Map 2: Northeast corner of Kodiak Island, showing Woody, Long, and Spruce Islands. Kodiak City is located at crosshatch in lower center.

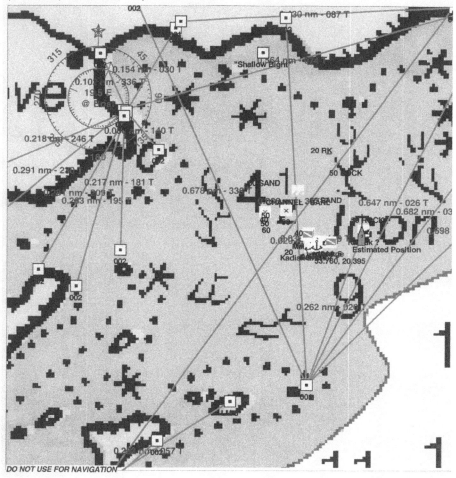

MARMOT BAY AND KUPREANOF STRAIT - 1 : 41,168
(NOAA Chart) Chart #16594_1 - Depth Units: FATHOMS

DO NOT USE FOR NAVIGATION

Map 3: After solving the puzzle of Arkhimandritov's bearings, I drew these lines connecting them. The estimated position of the *Kad'yak* is shown as a boat icon in center. Its actual position is shown as an anchor icon. I assumed that Captain Arkhimandritov took his final bearings to the ship from the rock at lower middle. They were probably taken from the small islet to the left of it, in which case, the bearings would have crossed the actual position.

Map 4: Aerial photograph taken from Mark Blakeslee's plane. Cross shows location of the *Kad'yak*.

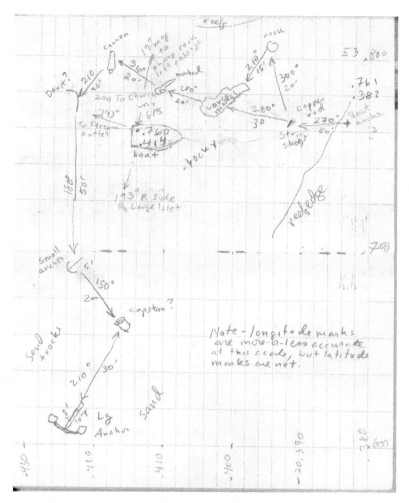

Map 5. My map to the major artifacts at the *Kad'yak* site, as estimated during dives made on July 27, 2003. Relative positions and distance estimates were confirmed during the survey conducted in July 2004. Notes in margin are latitude and longitude coordinates.

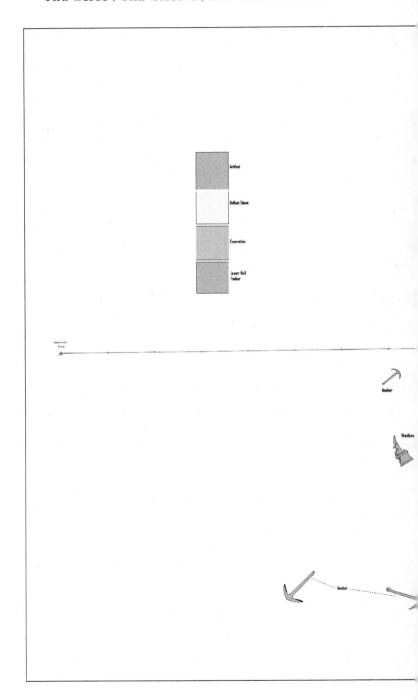

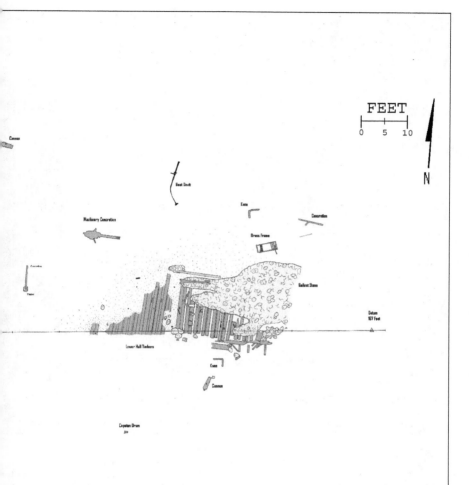

Map 6. Map of the *Kad'yak* wreck Site 1, showing locations of significant artifacts. Horizontal line is the baseline from which all artifacts were measured. The largest section consists of ballast stones and exposed wooden ship frames. From Cantelas et al. 2005. Item labeled Capstan Drum is the *Kad'yak* artifact

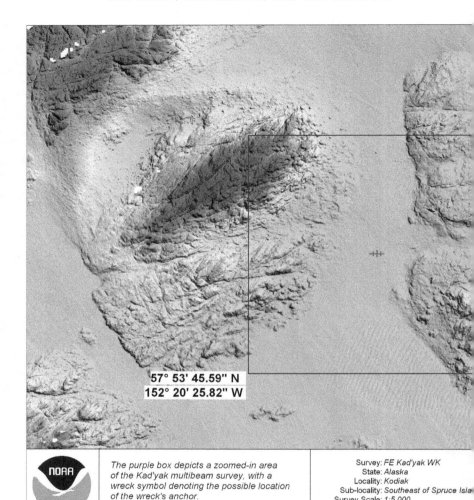

57° 53' 45.59" N
152° 20' 25.82" W

The purple box depicts a zoomed-in area of the Kad'yak multibeam survey, with a wreck symbol denoting the possible location of the wreck's anchor.

Survey: *FE Kad'yak WK*
State: *Alaska*
Locality: *Kodiak*
Sub-locality: *Southeast of Spruce Isla*
Survey Scale: *1:5,000*

Map 7. Multibeam survey of Icon Bay made by NOAA R/V *Rainier* in July 2004. Crosses show rough location of the *Kad'yak* wreck. Small bumps to right are the ballast pile, and small bump below it is probably the capstan. Parallel lines in lower center of figure are sand waves. (*Courtesy of NOAA.*)

7° 53' 48.17" N
52° 20' 20.52" W

Sounding Units: *Fathoms*
Sounding Datum: *MLLW*
Horizontal Datum: *NAD 83*
Projection: *UTM 5*
Scale Factor: *0.9996*

NOAA Ship RAINIER
CDR John W. Humphrey
Commanding

July 28, 2004

APPENDICES

APPENDIX A.
CAPTAIN'S LOG

Captain Illarion Arkhimandritov log surveyed the south end of Spruce Island, on July 12–14, 1860, and recorded the following information in his log. Following is a translation of his log by Katherine L. Arndt (in Italic) and my interpretation of their meaning. Notes are my own [bracketed remarks are my explanations]. In order to interpret the journal notes, I had to make three assumptions about the compass bearings:

1. Arkhimandritov did not write his bearings using 360-degree notation. Instead, he wrote them in quadrants, like "5 degrees SW," "45 degrees SE," "NE 35," and "NW 25." Tebenkov (1852) used similar notation in his Atlas, so it must have been common practice among Russian navigators (and maybe others?) at the time. No mention of this technique appears in Bowditch. I have interpreted these as bearings away from N or S. i.e, 5 SW means 5 degrees W of S (185 magnetic); NE 35 means 35 degrees E of N (mag 35), etc. That is the only way they make sense to me.

2. I assumed that all these bearings are magnetic and must be corrected to true bearings prior to use.

3. Magnetic variation at Kodiak is currently 19.4 degrees east, but varies annually. Applying the current rate of variation back 140 years gives a ridiculous result, so variation cannot be constant. Through an iterative process, I have determined that the magnetic variation was 22 degrees in 1860. Using this number, certain bearings to known landmarks are dead on.

At 6 o'clock in the morning, I began the description of this island from the SE end toward the N side. Bearings from this promontory: the promontory of this shore toward a narrow strait B SW 66 at a distance of 3/10 mile.

1. Doesn't specify what promontory. Most likely spot is a bluff at SE end of island called South Point (SP).
2. Using the assumptions stated above, SW 66 becomes 180 +66 + 22 = 268. This bearing is toward a small island almost due west of SP, and a channel is 0.3 mi away.

The right promontory of Monashkina (Monashka?) Bay is SW 5. Near this are 4 submerged rocks, the distance to a sea stack (?) is 3/10 mile.

3. SW 5 becomes 207 degrees (as above). This is a direct bearing to the bluff on the right side of Monashka Bay. Cannot identify the sea stack.

Melnichnyi point is SE 45.

4. This is a critical bearing. The translator interpreted this as Miller Point, but the bearing (SE 45 is 45 E of S, plus 22 = 157) is almost a direct shot to Spruce Cape, at 158. It is not close to (what we now call) Miller Point, at Fort Abercrombie, which would be 168, or SE 33. I assumed this bearing is to Spruce Cape, but am puzzled about it. Is this Arkhimandritov's mistake, or mine?
5. Lydia Black gave me a map drawn by Arkhimandritov in 1848 showing Mys Menichnyi (Mill Cape) at what we now call Spruce Cape. This vindicates conclusion #4 above. Apparently Miller Point was moved between his survey of 1848, and subsequent publication of his map (as Tebenkov's Atlas) dated 1849 (but actually published in 1852). When he resurveyed in 1860, he still thought of Spruce Cape as Miller Point.

The east promontory of an islet is NE 27; near this promontory are many submerged rocks all around at a distance of about 2 mile.

6. If taken from the same location as 1 and 2, this bearing (49° true) would be looking into the woods. However, if his boat moved slightly east, by about 0.1 mile, it is a bearing to, and description of the largest of three islets in Icon Bay.

The bearing toward the first landmark is NE 14.5°.

7. From the start point, this would be a course over land, through the woods. However, if he started from the same point as in #6, then it is a course along the shoreline. NOTE: Lydia Black says Arkhimandritov traveled from South Point to the NW along the S shore of Spruce Island, and did not survey Monk's Lagoon. However, all of the first eight bearings are to the NE. I think she is wrong and has not read these in detail.

On this course I moved toward the first landmark [what is it? A bluff?], the shore is steep, it slightly juts out from the island, there are rocks at 1/10 mile [offshore?]. The bearing of the second landmark is NE 35. On this course I moved toward this landmark; after 1/10 mile the mouth of a channel between the shore and an islet, the shore sinuous and rocky, is this landmark: the promontory of the first islet is between this landmark and the next or 3rd sea stack and at low tide it goes dry."

8. This describes a bearing (57°) along the shoreline and a channel (at 0.1 mi) at the tip of a small peninsula. A series of three islets or sea stacks are in a line just NE of here.

The bearing of the promontory of the second islet., i.e., the third landmark, is the same [NE 35°?].

9. There is a second small islet at exactly this bearing (57) from the channel mouth. Is this the third Landmark?

I moved toward this landmark, the island is 1/4 mile in circumference, it is steep, and there are rocks all around.

10. This is a good description of the first, largest islet (not the landmark, or second islet). It could only be reached on foot at low tide.

Bearings from this landmark: the right promontory of Selenie Bay is NE 23; the 1st isthmus from it is NE 14, the second isthmus is NE 4, the right promontory of a [notch or bight?] is NW 25, the left is NW 45.5, the topmast of the bark Kadiak is NW 2;

11. This is a critical landmark, as the Kadiak and many other points are described from it, but where exactly is it? What is meant by Selenie Bay? According to Lydia Black, Selenye means settlement. There were several "settler bays" around Spruce Island, including one marked on a chart at about the position of Icon Bay. If all these bearings are calculated correctly from one point, they should all point back to the landmark. "the right promontory of Selenie Bay" ...must be the outermost islet NE of Icon Bay, Ostrof Point. "the 1st isthmus...[and]...the second isthmus" must be shallow reefs connecting two islands to the shoreline, that would be exposed land (isthmus) at low tide. "right [and left] promontory of a [notch or bight?]..." must be a small bight in the northern shoreline of Icon Bay, but the bearings are off by a few degrees. These landmarks only line up if the starting point is the outermost or third sea stack, although it could have been part of the second islet prior to the 1964 earthquake.

On this course I moved toward the 4th landmark. From the third landmark I went to the left promontory of the new settlement; there are many submerged rocks and submerged creeks (?) Shallow water (?) And a small promontory with a small peninsula; there is a sea stack in the bight

12. When he writes "on this course I moved," I think he means "I moved along the following course," not referring to the previously mentioned bearing (NW 2). I do not know where the "new settlement" or "old village" were, but other landmarks can be used to define these locations. The sea stack may be the one on the left side of

what we call Monk's Lagoon (Chasovnia Cove). The left promontory
is not the fourth landmark (see #15 below) but is probably just a point
along the shore where he took more bearings; as follows...

*Bearings: the 4th landmark is NE 52; the right promontory of the village bight is
SW 28, the first promontory on the right shore is SW 44, the left promontory of
the old village is SW 7; the islet is SE 7; the direction of the shallow creek is SE 21;*

13. The fourth landmark is probably Ostrof point (see #20 below).
Bearing the reverse course of 52 (adjusted, as above) from there
lines up a point on the left side of Monk's Lagoon (point ML). Could
this be the "left promontory of the new settlement"? Bearings from
that location SW 28 and SW 7 define the right and left shorelines
of a small bay (the "old village" site?), though the bearings would
have to have been taken across land, where there may now be trees.
However, this interpretation is somewhat questionable because SW
44 and SE 21 do not line up with anything. The only marked creek (in
the center of Monk's Lagoon) is at NW 44 magnetic.

On this course I moved to the left promontory of the old village.

14. Again this term probably refers to the following text, not the previous
bearing. The punctuation is confusing because it follows a ; so it
appears to be part of the previous sentence.

*The shore is straight as far as the right shore of a small promontory and beyond it
is the old village of the Aleuts, in the bight are many rocks and one sea stack.*

15. This passage is very confusing. The shoreline here is very irregular.
To reach the location described above from the second or third
landmark would require a long, circuitous hike around two small
bays. However, he was traveling in a three-person baidarka, so could
go in a straight line, across the water. But if the old village site is
SW of the previous position (a conclusion supported by the previous
paragraph), it is opposite of the direction to the fourth landmark,
which is to the NE.

Bearings; the islet is SE 62, the distance to it is 1/10 mile;

16. If these bearings were also taken from point ML, they point to the
"sea stack in the bight" described in #12 above.

From it to the inlet of the shallow creek is 2 mile; the creek is almost dry at low tide.

17. From the sea stack it is actually 0.33 mile [but could have looked like
2 miles] to the creek on the map, which drains a bog and pond behind
the beach at Monk's Lagoon. When the tide is out, the creek outflow
sinks into the sand before it reaches the sea water, so could look like it
was "almost dry."

*From the left promontory of the new village I went to the right [promontory? Or just
a direction?] on a course NE 8. A shallow bight and many rocks, steep and sinuous.*

18. This bearing points to the bluff on the right side of Monk's lagoon. He

could not walk NE 8 from point ML, but could go by baidarka (across the water). Between this point and the next promontory, there is a shallow bight, the same one referred to in # 11 above.

From this promontory I went toward the 4th landmark on a course NE 65 (87). The shore is steep for 1/10 mile, has many rocks, and protrudes toward an isthmus; this promontory [the 4th landmark?] forms 2 small peninsulas, near the closer are 2 sea stacks and many rocks; there is a submerged rock at 3/10 (?) mile. Bearings [from this landmark?] the right promontory of Monaskha Bay is SW 14, Miller point [translator's interpretation; it is more likely he was describing Spruce Cape] is SE 26, the NE promontory of Long Island is SE 32.

19. The first sentence accurately describes the course from the right side of Monk's Lagoon to Ostrof Point.

20. Here are three bearings from the same location. The description of peninsulas and sea stacks describes the promontory nearest to Ostrof Point. From here, two of the bearings line up perfectly with Monashka Bay and Spruce Cape (Miller Point). The bearing to Long Island hits it about in the middle, rather than at the NE end. Therefore the fourth landmark must be at this location.

APPENDIX B.
NEWS RELEASES

Appendix B.1

"Breaking News—Shipwreck Found" *Kodiak Daily Mirror,* Friday July 23, 2003.

By Adam Lesh, Managing Editor.

Scientists today announced the discovery of what they believe to be wreckage of the Russian barkentine [sic] *Kadi'ak* which sank in 1860 in shallow waters near Spruce Island.

"We have discovered what we believe to be a significant archaeological site," said Dr. Brad Stevens, who spearheaded the project.

If the scientists are correct that their discovery is the remains of the *Kadi'ak,* then that makes it "the first shipwreck found from the Russian American colonial period, which makes it the oldest shipwreck found in Alaska."

The *Kadi'ak,* piloted by Capt. Illarion Ivanovich Arkhimandritov, was a Russian American Company ship, based out of Sitka but home-ported in Kodiak. The *Kadi'ak* transported ice from Kodiak to San Francisco.

"On March 30, 1860, it was carrying a load of ice, and after leaving the harbor, struck a rock," Stevens said.

The ship's cargo of ice, being buoyant, kept the ship afloat for four days. It finally sank near Spruce Island.

According to Stevens, the ship is surrounded in mystery. Legend says that when Arkhimandritov sailed the *Kadi'ak* the first time, the wife of the Russian governor of Alaska asked him to say a prayer in the chapel on Spruce Island near the place where St. Herman was buried. Arkhimandritov did not do so.

"When it sank, it came to rest with the mast and at least one spar forming a cross," said Stevens.

Stevens began the project when he received the translated journals of Arkhimandritov from Anchorage-based archaeologist Mike Yarborough.

The journals contained some bearings for the shipwreck, but due to differing methods for obtaining bearings, Stevens could not discern the resting place of the *Kadi'ak.*

However, he was able to discern one bearing in Monashka Bay, and another that Arkhimandritov called Miller Point.

Stevens discovered that the point called Miller Point in the journals was actually Spruce Cape when he saw a map from the Russian period, shown to him by Dr. Lydia Black. He was then able to discern the location of the ship.

Also involved in the discovery were Steve Lloyd; Stephan Quinth, a filmmaker from Sweden and his assistant; Josh Lewis, who lent his boat, the *Melmar* for use in the project; Dave McMahan, an archaeologist for the Alaska Department of Natural Resources; Bill Donaldson; and Verlin Pherson, a diver.

The Project was funded by donations from Kodiak Historical Society members.

Stevens stresses that the site is in state waters, under the protection of the State of Alaska, and protected by state statutes.

"We can't have other people going in and disturbing the site," said Stevens.

Stevens said the site will probable [sic] qualify for inclusion in the National Historic Register.

Mirror writer Justin Blomsness contributed to this story.

APPENDICES

Kodiak considers its treasure-trove
SHIPWRECKS: Diving enthusiast touts island's untapped assets.

The Associated Press (Published: December 9, 2003)

KODIAK – Disasters from Kodiak's past could turn into blessings for its future, according to a maritime lawyer and shipwreck diving enthusiast.

"You guys have a great resource in the maritime history of this island," Peter Hess told an audience Sunday at Kodiak College.

About 40 people attended the lecture by Hess, of Wilmington, Del., sponsored in part by the Kodiak Maritime Museum.

Audience members heard stories of silver, gold and jewels salvaged in recent years from wrecks dating to the days of the Spanish galleons. Hess recalled his excitement at seeing real treasure chests bursting with pieces of eight.

He said advances in scuba diving technology will make sunken ships around Kodiak more accessible. Hess foresees a time when diving could join fishing and hunting as a local economic asset.

"I think you have the beginnings of a new industry," he said.

Among the most interesting and accessible local wrecks is the *Kadiak*, a Russian-American merchant barque carrying a cargo of ice that sank in about 80 feet of water off Spruce Island in 1860, making it the oldest wreck found in Alaska.

Hess visited the *Kadiak* on Saturday and was able to touch one of the cannons it carried for protection.

"This is a world-class site that will bring people in from the East Coast and all over," he said.

Hess' underwater partners were Steve Lloyd of Anchorage, who discovered the wreck's exact location last summer, Josh Lewis of Kodiak and Andrey Nikolaev, a recreational diver from Sakhalin Island, Russia.

Out of the water, Hess spends much of his professional time navigating the maelstrom of competing claims, precedents and jurisdictions that result from discoveries.

"Once something is found, the first thing that's asked is 'Who owns it?' " he said.

Assuming that a ship's owners have abandoned a wreck, the basic principle of salvage law is simple: finders keepers. That still has validity in the maritime context, Hess said.

He participated in the historic salvage of *Nuestra Senora de Atocha*, a galleon that sank in 1622 in the Florida Keys and was discovered in 1985 by salvager Mel Fisher. After years of litigation, some investors in Fisher's operation earned a return of 6,000 percent.

Hess blames ill-conceived government regulation for endangering the salvage industry by restricting access. Noting that treasure hunters have a bad name, he said regulation should encourage salvagers to do the right thing by being responsible archaeologists.

Without the expectation of some compensation, historically significant wrecks will disintegrate in the harsh underwater environment or fall into the hands of looters, he said.

In the case of the *Kadiak*, Hess said, he hopes the people of Ouzinkie will have substantial say in its future because of their historical connection.

"They ought to be the ones deciding what the disposition of this wreck ought to be," he said.

He recommended the Kodiak Maritime Museum as an ideal repository.

"When you get involved with a project like this, you realize the real treasure is the history," he said.

Hess participated in dives on the wreck of the *Monitor*, the Union's iron-clad ship famous for its battle with the Virginia, more commonly known as the *Merrimac*, in the Civil War. He said his excitement at finding an ordinary dinner plate there at least equaled the thrill of lifting silver ingots out of a treasure ship.

APPENDICES

Letter to Editor, *Kodiak Daily Mirror*, December 2003, by Bradley G. Stevens

Dear Editor,

Ever since the discovery of the *Kad'yak* shipwreck, I have tried to stay above the fracas that has been roiling around it. However, in response to recent events reported by the *KDM* (December 9, 2003), I wish to set the facts straight.

Josh Lewis, a Kodiak teacher, and Steve Lloyd, owner of an Anchorage bookstore (Title Wave), have claimed to be the discoverers of the *Kad'yak*. Apparently, neither the veracity of their statements, nor the legality of their actions was ever questioned by your reporter. Over ten years ago, translations of Captain Arkhimandritof's logs were provided to me by archaeologist Mike Yarborough. I used those translations, plus a map provided by Dr. Lydia Black, to determine the probable location of the shipwreck. Do Lewis and Lloyd have either of these items?

In 2002, I began working with Dave McMahan, state archaeologist for Alaska, and Dr. Timothy Runyan, Director of the East Carolina University (ECU) Department of Maritime Studies. Dr. Runyan and his associates are highly qualified marine archaeologists and historians. Together, we submitted a proposal to NOAA for funds to search for the *Kad'yak*. When that application was not funded, we began planning a preliminary diving survey for 2003.

In May 2003, I discussed our plans to search for the *Kad'yak* with Josh Lewis, who offered the use of his boat, and invited Steve Lloyd to accompany us. Other divers who assisted us were filmmaker Stefan Quinth, Bill Donaldson and Verlin Pherson. The Kodiak Historical Society provided financial support for our efforts. Dave McMahan also participated in the search, in lieu of providing the necessary permit from the State of Alaska, thus making our survey legal.

During two days of diving in July 2003, we found the shipwreck within 100 yards of my estimated position. Our search was aided by the use of a magnetometer operated by Steve Lloyd. On my last dive, I photographed, measured, and mapped the wreck site. All of the divers, including Lewis and Lloyd, were instrumental in finding the shipwreck, but we would have found it even without their help. Would they have found it without mine? You decide.

Before diving, Dave McMahan explicitly told all the divers that the *Kad'yak* wreck was in State waters, and any artifacts found there belonged to the people of Alaska, represented by the State. This was clearly understood by everyone. During the dives, Steve Lloyd and I both recovered small artifacts from the wreck site. Most of these were turned over to Dave McMahan at the time of discovery, but two items were apparently kept by Steve Lloyd. Requests for return of the artifacts were initially ignored, but they were finally returned

THE SHIP, THE SAINT, AND THE SAILOR

three months later, after the Alaska Department of Natural Resources and State Attorney General's office threatened legal action. Why did they keep these artifacts? Did they plan to file a private salvage claim on the *Kad'yak*? Do they still have other artifacts in their possession?

In November 2003, the ECU team submitted another grant proposal to NOAA, including $8000 to develop a shipwreck curriculum for Kodiak students. The Kodiak Historical Society, Baranov Musuem, Alutiiq Museum, and the Kodiak Maritime Museum all support this project. The State of Alaska has issued us an archaeological research permit, in recognition of our qualifications. Lewis and Lloyd also applied for a permit, but their request was denied. Not only did they lack archeological expertise, but their past actions were highly questionable.

Now, apparently, Lewis and Lloyd have teamed up with Peter E. Hess, a Delaware lawyer, to represent their claims to the *Kad'yak*. The *KDM* story indicates that they visited the *Kad'yak* site and handled or disturbed some of the artifacts. If so, they have interfered with the State Archaeology Permit issued to ECU, and may have violated State law as well. They appear to be promoting the *Kad'yak* site as a recreational dive opportunity that will "bring people in from all over". As a recreational diver, I enjoy wreck diving too, but the *Kad'yak* is not an intact wreck. Artifacts are scattered over a wide area, and most are buried in the sand. If I came from some distance to see it, I would be greatly disappointed.

They claim that attracting divers to Kodiak would benefit the community. Perhaps it would sell a few hotel rooms and meals, but I think that the major benefactors would be a few self-interested parties. On the other hand, a proper archaeological survey of the site, and eventual recovery, conservation, and display of the artifacts would benefit a much larger segment of the community, as well as educate many tourists, and the world at large, about Kodiak history.

Alaska's State Historic Preservation Office has determined that the *Kad'yak* wreck site meets eligibility requirements for the National Register of Historic Places. This should be reason enough to protect it from pilferage. It would be a great shame if small artifacts ended up in the pockets of recreational divers, instead of a local museum, where they belong. Without proper conservation, that is exactly what will happen.

Hess claims that the rule of law is "finders keepers", but both federal and state laws declare that submerged cultural resources belong, not to the finder, but to the public, represented by the State. Hess also said "the people of Ouzinkie ought to be the ones deciding the disposition of the wreck". I agree that they should participate in the process, because they are linked to the *Kad'yak* by it's location. Both Tim Runyan and I have invited the Ouzinkie Native Corporation to participate, and received permission to conduct surveys on their property.

We look forward to their collaboration.

But this statement seems like an attempt to create division among the local community, because all the people of Kodiak should be involved. The *Kad'yak* was named for Kodiak, sailed from Kodiak, carried a cargo from Woody Island, and sank near Kodiak, eventually becoming one with the island for which it was named. Rather than arguing over who should decide its fate, the people of Kodiak Island, its villages, and the local museums should be united in this effort to study and conserve it. Bringing up the artifacts to the light of day will bring our history to light as well. Who deserves more to see that history? Some rich scuba divers from out of town? Or the people of Kodiak, Ouzinkie, and the rest of the State?

Other writings by Peter Hess and his associates make it clear they believe that recreational divers should have rights to visit and salvage wrecks, and that the government and archaeologists should get out of the way. While I agree that wreck diving is a fun and popular sport, it should not take precedence over archaeological research on wrecks of historical significance, such as the *Kad'yak*. The government is not going to take anything away from Kodiak, or the *Kad'yak* site. To the contrary, we believe the government should provide the funds that will allow Kodiak to benefit from research and recovery of the shipwreck. The additional boon of $8000 to financially-strapped local schools is not something to be scoffed at.

I am a biologist, not an archaeologist or historian. But I have a great deal of professional experience and training in underwater research and exploration. I personally stand to gain nothing from this endeavor except headaches. I set out to find the *Kad'yak*, and am overjoyed to have achieved that goal, with help from all the participants. My only motivation was to enhance Kodiak's history, and to support the real archaeologists in their efforts. In the end, who should get credit for the discovery is of little importance. There is no treasure to be found. The only real treasure is history, and those who should benefit from it are not just amateur wreck divers, but all the people of Kodiak.

The *Kad'yak* is the oldest shipwreck of historical significance discovered in Alaska, and the only one from the Russian colonial period ever discovered. How we treat this discovery will set a precedent for all other Alaskan shipwrecks that may be found in the future. We have a chance, and an obligation, to do the right thing. Clearly, we should not squander that opportunity.

Sincerely,

Bradley G. Stevens, Ph.D.

APPENDIX B.4

Alaska's maritime history belongs to all of us. By David McMahan

Letter to Editor, *Kodiak Daily Mirror*, December 2003.

Dear Editor:

I am writing in response to an article entitled "Kodiak considers its treasure-trove" that appeared in the *Kodiak Daily Mirror* and, through syndication, in the *Anchorage Daily News* on 12/9/03. The article discusses a recent public talk given by Delaware maritime lawyer Peter Hess. Particular attention was given to a discussion of the recently discovered wreckage of the Russian-American merchant barque *Kad'yak*, which sank in 1860. I have not met Peter Hess, but know him through his writings in opposition to international initiatives to protect historic shipwrecks through UNESCO. As an archaeologist for the state, I have been involved with historic shipwreck issues for several years. I participated in the dives which resulted in the discovery of the *Kad'yak* in July 2003 and facilitated the collaboration of researchers that led up to the discovery. I would like to offer clarification on some of the issues surrounding the *Kad'yak* and Alaska's submerged historic resources in general.

Anchorage archaeologist Mike Yarborough and several associates began researching the *Kad'yak* in 1979, assembling and translating archival materials that eventually pinpointed the location of the wreck site within about 300 feet. It is important to note that no agency or organization within Alaska has an underwater archaeology program, yet our waters abound with submerged archaeological sites that can add to our understanding of Alaska's history. Documentation of these sites requires the special skills of archaeologists trained in underwater techniques. For this reason, I facilitated a collaborative relationship between the State of Alaska, East Carolina University (ECU) maritime history professor Dr. Tim Runyan and his associates, Mike Yarborough, and NOAA biologist Dr. Brad Stevens. ECU has one of the best known underwater archaeology programs in the U.S. and is able to provide expertise in maritime history and underwater archaeology that is absent in Alaska. The search that resulted in the discovery of the *Kad'yak* last summer was a scaled-down effort initiated by Dr. Stevens in collaboration with the state. Shoreline Adventures, whose help I gratefully acknowledge, assisted in the effort at the invitation of Dr. Stevens through dives and the use of their boat and equipment. I also acknowledge assistance by the Director and Board of the Baranof Museum, who generously agreed to subsidize boat expenses.

Following the review of a comprehensive research design, the Alaska Department of Natural Resources issued a permit to ECU to document and map the wreck site. The *Kad'yak* consists not of an intact hull, but rather a constellation of durable fittings, implements, and accoutrements that became imbedded in the sea floor after the wooden hull decayed. The permit issued to ECU allows for mapping and limited artifact removal, but does not allow for large scale artifact recovery or excavation. All artifacts from the site are considered state property. They will remain in Alaska, or will be returned to the state following specialized conservation treatments at an outside laboratory. Since the inception of the project, the collaborators have consulted with and received encouragement from local Kodiak museums. In their grant proposal for work at the *Kad'yak* site, ECU has requested funding for community involvement. We all believe that any recovered artifacts belong to the residents of our state and should ultimately be exhibited in Kodiak.

Unfortunately, the recovery of submerged artifacts is the easy part of any underwater recovery effort. The location and position of each artifact must be precisely recorded and mapped if meaningful information is to be realized. To indiscriminantly collect artifacts without recording context robs us of available knowledge, much like discarding a page from a one-of-a-kind book. Most classes of artifacts quickly begin to deteriorate upon removal from saltwater if not subjected to intensive, and often expensive, conservation treatments. The budget and effort for conservation often far exceeds that of recovery – and once the thrill of discovery is gone, so is the incentive and funding to preserve the collection. Our position is that submerged artifacts should remain in place, in equilibrium with their surroundings, until a specific plan and budget are in place for removal, conservation, permanent curation, and public exhibition.

The state law (11 AAC 16.030) under which archaeology field permits are issued requires that applicants meet accepted professional standards. Permits are issued to the first qualified applicant and allow for the exclusion of others from the same site for the duration of the permit. With regard to the *Kad'yak*, the Kodiak troopers were notified, following consultation with the State Attorney General's Office, that the site is presently off limits to unauthorized divers due to protection concerns. This was publicized in the *Kodiak Daily Mirror*. Consequently, *Daily Mirror* coverage that Peter Hess, Steve Lloyd, Josh Lewis, and Andrey Nikolaev made unauthorized dives on the site and handled artifacts is disheartening.

While the law of shipwrecks is sometimes complicated, it is important to understand the basic legal and ethical framework which governs the way in which these resources are treated. Salvage law, founded on ancient Roman legal principles, essentially allows for "saving property that is in peril at sea, and returning it to the owner for a reward." Unless the owner has agreed, the salvor's claim for a reward is generally against the property in rem (i.e., against the ship herself), which requires that the ship be "arrested" under admiralty process. Salvage claims are considered in federal admiralty courts. Salvage law rewards the economically efficient recovery of commercially valuable objects but promotes the unscientific destruction of historic wrecks and the permanent loss of archaeological information. Shipwrecks have often been compared to time capsules because unanticipated sinkings often result in a broad spectrum of artifacts from a confined time period. The U.S. Congress recognized the importance of protecting shipwrecks as historic resources by passing the "1987 Abandoned Shipwreck Act" (ASA). Under the ASA, the federal government took title to all wrecks within three miles of the U.S. coast which are abandoned, and either (a) embedded in the sea bed or coral formation or (b) eligible for the National Register of Historic Places (i.e., historic). The federal government then transferred title to coastal states with guidelines on how to manage the wrecks.

Alaska state law (AS 41.35), reinforced by the ASA, asserts ownership claims and protection measures on all historic resources on/in state lands, including state tidelands and state submerged lands – i.e., within three miles of the coast. As a rule of thumb "historic" properties are more than 50 years old, although more recent properties can achieve historic eligibility if they are found to be of exceptional significance. Alaska's State Historic Preservation Office has determined the *Kad'yak* eligible for inclusion on the National Register of Historic Places.

Existing laws adequately protect some wreck sites, while many others are lost in the haze of unclear legal standards and enforcement inconsistencies. Historic shipwrecks are fragile, non-renewable public resources. It was never the intent of the Abandoned Shipwrecks Act to systematically exclude recreational divers from the thrill of exploring shipwrecks, although it does allow for exclusion from certain sites that are considered particularly fragile or threatened. The State of Alaska, unlike coastal states in more populated areas, does not have a program for the archaeological study of submerged resources. The most productive means for us to investigate and record information from these sites is through collaboration either with federal agencies such as NOAA, with universities, or with recreational dive groups and local dive shops. The world's largest recreational diver certification organization, PADI, teaches its members to "take only pictures - leave only bubbles." We embrace this ethic, and welcome collaborative opportunities for joint exploration and outreach.

Finally, it is important to note that the *Kad'yak* was not laden with gold – but with ice packed in sawdust and bound for iceboxes and cocktail glasses in the bustling city of San Francisco. The treasure-trove of the *Kad'yak* is not in her cargo, but in the wealth of historic information that can be gathered through scientific methods and shared through the development of museum exhibits and publications.

APPENDIX C.
DESCRIPTION AND DISPOSITION OF THE *KAD'YAK* ARTIFACTS

During the 2003 discovery dives, a number of small pieces of copper sheathing, two drift pins and a metal bracket were removed from the site by Steve Lloyd without adequate provenance (so no record was made of where they were found relative to the remainder of the wreck). As a result, they don't contribute much to the archaeological analysis. That is what happens when untrained amateurs remove artifacts from archeological sites. That it happened on my watch, so to speak, is somewhat embarrassing. I shouldn't have let it happen, but at that time we were all amateurs.

In contrast, during the 2004 survey, all visible artifacts were geolocated, documented, and recorded, and a few of them were recovered for conservation. At the time, we could only speculate on what they were and what their significance was. Analyses, detailed drawings, and descriptions were made, and some items were sent out for laboratory analysis. Most of the larger artifacts were left on the site, whereas smaller items that were deemed subject to looting or loss were recovered for preservation.

Detailed descriptions and photographs of the wreck and artifacts can be found in the comprehensive report of the 2004 *Kad'yak* survey expedition, produced for NOAA and the National Science foundation by Frank Cantelas, Tim Runyan, Evgenia Anichenko, and Jason Rogers (Cantelas et al., 2005), and in another document (Anichenko and Rogers, 2007), that summarizes the value and meaning of the *Kad'yak*'s history in both English and Russian text.

We separated the wreck site into two sites. Site 1 includes the major portion of the wreckage, consisting of the ballast pile, anchors, windlass, cannons, and other materials found on a sand and gravel bottom between two rock reefs, and spread to the west (Map 6). Numerous wood frames, probably made from white oak, lie beneath the ballast pile, in roughly a N-S direction, which would have been athwart-ship. Most of these have been exposed by water funneling through the narrow passage between two nearby reefs, and have been extensively damaged by sand scouring and shipworms. There are probably more wood frames to the west that are still covered by sand. There are also many bronze drift pins and pieces of scattered metal around. The metal sheathing of these pieces was determined by lab tests to be Muntz metal, a form of brass with a composition of 62 percent copper and 28 percent zinc.

One hundred and thirty feet northeast of Site 1 lies Site 2, where I had found pieces of metal in 2003. Only one day was spent exploring this site in 2004, so much still remains to be learned about it. It consists of a linear debris field containing many unidentified elements, most of which probably came from the stern section of the ship. These include pieces of iron pipe, drift pins, and chain, which may have been parts of the steering mechanism, and much of which is cemented together. Many of these items were arrayed across the reef face and down onto the sand. Gudgeons and a pintle were found at the base of the reef.

Below are descriptions of some of the major artifacts found on site, plus smaller items that were recovered for conservation. Positions of most of these are depicted on Map 6.

WINDLASS — One of the largest objects, and one of the first to be found, is a portion of the ship's windlass. The windlass was used to haul up the heavy anchors, tighten mooring lines, and occasionally to haul up sails. It was usually operated by a team of men pushing handles while walking in a circular motion. The object we found was an iron tube approximately 2 feet wide and 3 feet long, with an octagonal core that was probably made to fit around a wooden windlass barrel. Several sections of heavy iron chain are cemented to it. Chain links (rodes) for anchors came into use only after 1800 and were common by mid-century, and an anchor and chain cable were listed as cargo on the *Kad'yak* during a voyage in 1859. Nearby the windlass we found a heavy, rectangular iron frame, about two by three feet. Although its shape suggests it may have been a hatch frame, it is probably too small for that purpose

ANCHORS — Three anchors were found, two of which were bower anchors. The largest is approximately 8 feet long, with straight arms, similar to those produced in the early 1800s. A smaller, auxiliary bower anchor has round arms, of a kind produced in the 1840s. Both of these would have been used to anchor the ship in place. The smallest anchor appears to be a kedge anchor. All three are in close proximity and therefore probably indicate where the bow of the ship came to rest. Anchors may have been traded among ships of the RAC, accounting for the different designs found.

CANNONS — Two cannons were found at the site, which we later labeled "north" or "south" according to their position relative to the centerline. The north cannon was the first found, in 2003, and lay 43 feet north of the centerline. It is 3 feet long, with a bore of 3.6 inches, and probably fired a 6-pound cannonball. The south cannon was found in 2004 11 feet south of the centerline. It is 4 feet long, with the muzzle broken off. Both cannons are cast iron and typical of modified "carronades" that were used by Russian ships at the time. The historical record suggests there were six cannons aboard, so others may have rolled off when the ship heeled over. No cannonballs were ever found, other than the one that I "exploded".

SKYLIGHT GRATE — A coppery metal frame or grate was first observed during the dive made in February in 2004. It is approximately 4 by 2 feet, and

originally had twenty-four crossbars, of which only eight remain. A cabin
on the deck of the *Kad'yak* was removed after it arrived in Kodiak because it
proved not to be seaworthy. It may have been replaced with a skylight, with this
grate as its cover, protecting the glass.

WHEEL HUB — This is the item referred to as "the *Kad'yak* artifact" and
generated the greatest excitement. It was labeled by news reports as the Holy
Grail of the site, because it identified the ship. It consists of a wooden cylinder
about twelve inches long and seven inches in diameter. One end was covered
with a copper-alloy cap, bearing the Cyrillic letters *К О Д Я К Ъ* engraved in
the metal. The other end had a square hole, surrounded by an iron frame. It
seems likely that a metal rod was inserted into the square hole. The presence
of the ship's name on this item indicates its importance and suggests it may
have been the hub of the ship's wheel, from which wooden spokes would have
protruded.

BILGE PUMP — This is a long copper tube with lead fittings on each end. It would
have formed the major portion of the chamber for pumping water out of the bilge.
Bilge pumps were usually placed in pairs on either side of the keelson, in the
deepest part of the hull, which would have been near the main mast. The bottom
section of one bilge pump was found embedded in a timber frame next to the
ballast pile, indicating that location as the center of the ship. The tube was found
over 130 feet from the centerline, where it had been carried by water currents.

BRACKET – This item consists of cast copper alloy shaped into the form of a
hook, approximately 8 inches long and 4 inches high, bonded to a rectangular
base. Marks on the base suggest that a similar hook extended from the other
side. We're not sure of its purpose, but it looks to me as if it might have been
used for hanging lanterns.

VALVE – This item is a valve made of copper alloy, with portions of lead pipe
extending from it, and was probably attached to a water barrel (though I like to
imagine it protruding from the bottom of a rum barrel). Drinking water from a
lead pipe would not have been healthy though.

GUDGEONS AND PINTLE – The gudgeons and pintle are heavy metallic
parts found at Site 2, and were used to attach the rudder to the sternpost. The
gudgeon is a casting with a large hole in it, with straps through which pins
were inserted into the sternpost. The pintle is similarly shaped, but with a
cylindrical pin, that fits into the hole in the gudgeon. Pintles would have been
attached to the rudder, with their pins extending downward, and were held
in place by the weight of the rudder. One intact pair of pintle and gudgeon
was found, though the attachment straps were broken off of the latter. After

bringing them on board the *Big Valley*, we separated them, possibly for the first time since the rudder was mounted on the ship. Another gudgeon was found close by. These parts indicate that the stern of the ship, including most of the steering gear and chain, probably settled at this part of the reef.

Several other large metallic objects were located, including a boat davit and unidentified machinery; these were left in place after recording their positions. Small items that were found and preserved include unidentified metal pins and shafts, a barrel stave, and a piece of wood identified as western hemlock. The latter that may have been cargo, intended to be used as firewood, or may have been "dunnage" used to fill space below and around the ice cargo. Since hemlock is not native to Kodiak, it probably came from some other location. Items that were collected for preservation and analysis were treated differently, depending on their consistency. All items were soaked in fresh water for up to four months, with periodic changes. This was followed by soaking in a solution of sodium sesquicarbonate for five weeks. Items were then placed back into tap water with periodic changes until the concentration of salt was reduced below ten parts per million, or ppm (for comparison, seawater is 35 parts per thousand, or 35,000 ppm). The wheel hub was then sent off to Texas Agricultural and Mechanical University, where it was immersed in a polymer solution. Other items were soaked in acetone, followed by ethyl alcohol to remove water. Finally, they were air dried and given a coating of clear acrylic Krylon.

After final preservation, the hub and other artifacts were transferred to the Alaska State Museum in Juneau. They are now part of a maritime exhibit in a new museum building. Under state protocols, artifacts from state lands are permanently curated by the state museum system but are generally available on loan to smaller museums for special exhibits. While this is sometimes controversial or misunderstood in smaller communities, smaller museums generally do not have space or the resources to take care of archaeological collections permanently. It is to their benefit to borrow exhibit items on an "as needed" basis and have larger museums be responsible for curation and documentation. Although this arrangement supports those who argued that the artifacts would not be returned to Kodiak, these precious artifacts will now be preserved in better condition, and for a much wider audience, than would be possible in Kodiak, where none of the local museums or organizations have the resources for this purpose. It is my hope that these artifacts will eventually rotate around to Kodiak and other communities, for display and education.

ENDNOTES

1 Looking back on it from my current perspective, having written and submitted dozens of grants, I realized that, like any good grant writers, we needed to insert the accepted jargon of their particular branch of science into the text. To convince the people reviewing the grant (most of whom were scientists who would rather be doing our project than reviewing it), we had to make the proposal sound like the work of knowledgeable and credentialed professionals just like the reviewers, who were experts in our fields, knew exactly what we were doing, and could accomplish our goals with astounding success, under budget, and ahead of schedule; we had to make the reviewers see they would be complete buffoons if they did not agree with our reasoning and offer us the funding immediately. At least that's what we wanted them to think.

2 I laugh about this now. Every grant I submit now has to include a full year of student support, a month of my salary, benefits for both of us, and university overhead, all of which can easily add up to $50,000 before we even leave the dock! Add in some grants the equipment, supplies, travel, and cost of chartering a boat, and $100,000 doesn't look like much anymore. I rarely request less than that because it requires just as much paperwork to ask for $10,000 as $100,000, so why bother?

3 Much later, I learned that Josh had deliberately delayed picking us up in an effort to blunt my leadership role. Although we were not much worse for the wear, forcing us to commit to a long surface swim was a negligent, unnecessary, and unsafe thing to do. If this had been an official NOAA or university-sanctioned dive, as a trained dive supervisor, I would have insisted on sending the skiff to pick up the divers. Looking back, it was one more nagging piece of evidence that Josh and Steve were not working for the benefit of the team.

4 The reason Steve wanted to have our back-deck conversation recorded on video by Stefan wasn't just to record the discovery of the first artifacts from the wreck site. It was to prove that Steve had found them. I didn't know this until much later.

5 This item later turned out to be an important piece of evidence, marking the location of debris that we would later identify as the stern section of the ship that had broken away from the rest of the ship.

6 In retrospect, maybe I was too cautious about not giving out a few juicy details of the wreck site. That plus some photos would have made for a much better news story. But I had real concerns about other divers looting the site. What I didn't know was that the looters were among our midst.

7 Did I really believe that a few photographs would entice British pirates to fly over to Kodiak to steal a cannon? I'm sure they have plenty of their own.

8 In fact, I have built a similar system in a 5-gallon bucket to remove rust from sailboat hardware.

9 To this day, I don't know if the *KDM* published it or not, and I don't really care. I just needed to vent.

10 The show later aired as an episode of *History Detectives* on season two, episode nine.

11 Check it out at http://dnr.alaska.gov/parks/oha/projects/kadyak#discovery.

12 One thing I have learned from Tim: Not all artifacts are worth recovering. His case in point was a barrel they recovered from the *Queen Anne's Revenge* that contained nothing but nails. But once they had recovered it, they had to go through the process of conserving every nail. It made him think twice about what they recovered after that.

NOTES

I N THE FOLLOWING NOTES, MY text (placed in quotation marks), serves only to identify the sentence or paragraph to which I refer, and is not an actual quote. This is followed by the source of the information, which, in most cases, has been transcribed, condensed, or summarized.

Chapter 1

"Their course required them to make several tacks..." Tebenkov (1852: 57 in Pierce's translation) describes the difficulties of entering and exiting Paul's Harbor in Kodiak, noting that the western entrance was preferred.

"So although its exact dimensions were not recorded..." Anichenko and Rogers (2007: 15).

"...the barque was defined by its rigging..." Desmond (1919:123).

"Under the command of Captain Bahr..." Anichenko and Rogers (2007: 15).

"The *Kad'yak* first carried ice to San Francisco in 1857..." Ibid, p 23.

"Captain Arkhimandritov was well known..." Details of the life of Captain Arkhimandritov were described by Pierce (1990: 10).

"In September of 1842..." Tikhmenev (1978: 363) recounts the story of the Naslednik Alexsander.

"Prior to his last trip to Kodiak..." The story about Captain Arkhimandritov reneging on his promise is told by several sources, including Golovin (1862: 127), which may be the source for Pierce (1990: 11).

"The *Kad'yak* had hit a rock..." Multiple versions of the sinking are recounted by different authors. This account was reported in a dispatch from Sitka to the Russian American Company Headquarters on December 22, 1860 (cited as Morskoi Sbornic 1861:204-205). A translation of this document is included in Anichenko and Rogers (2007:27), and Cantelas et al (2005:24). The "definitive" version was reported by Tikhmenev (1978: 362) and briefly summarized by Pierce (1983: 65).

Chapter 2
"...the Alutiiq word *Kikh'tak*" Dr. Sven Haakanson, Director, Alutiiq Museum, pers. Commun., 2003.
"... where he died in December 1741." Black (2004: 48).
"The first permanent Russian settlement..." Tikhmenev (1978:20).
"soft gold..." Black (2004: 77).
"In 1792, the settlement at Three Saints Bay..." Black (2004: 141).
"Baranov even declared..." Tikhmenev (1978: 153).
"...Shelikhov's company was granted a charter..." Black (2004: 211). This was the first of three charters granted to the RAC. They were renewed for additional 20-year cycles in 1821 and 1841 (Tikhmenev, 1978: 244).
"Alexander Baranov arrived..." Black (2004: 107).
"As late as 1868..." Scammon (1874: 174).
"Hunting parties were usually organized..." Tikhmenev (1978:441).
"Russian relations with the Natives were complicated. ..." Black (2004: 257).
"Disagreements between Baranov and the Church..." Black (2004: 239).
"In 1805, the Russian colonies were visited..." Tikhmenev (1978: 92). Accounts of Rezanof's brief trip to Alaska differ greatly by author. Tikhmenev, as historian for the RAC treats him quite deferentially, giving him great credit for putting the RAC on an even keel and fixing many problems.
"In 1805, the Russian colonies were visited by Nikolai Petrovich Rezanof ..." Black (2004: 172). In contrast to Tikhmenev's account, Black describes Rezanof as a disruptive blowhard who claimed responsibility for many reforms and discoveries that he did not actually achieve. Among his other "achievements," in an attempt to establish trade with Spanish Mexico, he sailed into San Francisco uninvited, announced himself as "Commandant General" of Russian America, and at the age of 42, engaged himself to the 15-year-old daughter of the Spanish Commandant Arguello, a proposal which he probably never intended to keep. He also wrote many letters requesting the RAC to conscript serfs and criminals for service in Russian America, implored them to imprison Japanese laborers, and sent warships to invade Japan, in an ill-fated attempt to open their doors to trade. After returning to Okhotsk in 1807, he set out on a return trip to St. Petersburg overland during the winter, but died in his native Irkutsk. After stirring up a whirlwind of both expectation and anxiety in the Colonies, his efforts there had little lasting impact.
"He also found a champion in Lieutenant S. I Ianofskii ..." Pierce (1990: 166).
"Later in his life...protector to Anna Baranovna..." Black (2004: 132).
"One day there was a terrible earthquake." The story of Father Herman's "miracle" is related by Anichenko and Rogers (2007: 29).
"Father Herman died in 1836..." Black (2004:239), but Tikhmenev (1978:413) states that he died in 1827.
Chapter 3
The complete translation of Captain Arkhimandritov's Journal of the Spruce Island circumnavigation, as translated by Kathy Arndt, along with my interpretation of his bearings, landmarks, and their meanings, is included as Appendix A.
"The sinking of the *Kad'yak* was a great loss..." Anichenko and Rogers (2007: 27).
"It is strange..." quotation cited from Anichenko and Rogers (2007:27), and Cantelas et al (2005:24). Original source was a letter from New Arkhangelsk dated December 22, 1860, published in *Morskoi sbornik* (Maritime Digest) 1862.
Chapter 4
"all I could see were crabs..." B. Stevens et al. (1994).

Chapter 5
"The *Delta* was a workhorse for the marine biology research community." For an entertaining history of the *Delta* and its adventures, try to find a copy of Richard Slater's self-published biography (2012).

Chapter 9
"But in 1988, the Abandoned Shipwrecks Act (ASA) was passed by Congress..." 100th Congress of the United States, Abandoned Shipwreck Act of 1987.

"There were supposed to be four to six cannons on the ship..." Cantelas et al. (2005: 68).

Chapter 11
All the artifacts found were fully described by Cantelas et al. (2005: 43-67). Here I provide brief descriptions and explanations of items as they were found.

"The first things I saw were copper rods..." These were bronze drift pins, used like long rivets to hold the ship together. Most were 12 to 24 inches in length.

"A thick leather strap..." Despite being underwater for more than 140 years, this item did actually turn out to be a leather strap.

"The large mass was about 8 feet long..." This was the ballast pile, consisting of mostly granite boulders.

Chapter 12
"...the *Kad'yak* story was the front-page headline..." *Anchorage Daily News*, July 28, 2003. Appendix B2.

Chapter 13
"By this time, the RAC was bloated..." G. Stevens (1986A). The history of the Woody Island ice-cutting operation was summarized in a series of newspaper articles by Gary Stevens, a Professor of History at Kodiak College, University of Alaska, and long-time State Senator. For a more exhaustive history of the ice trade in New England and the US East Coast, see Weightman (2003).

"The market for furs in China..." Black (2004: 263).

"It was also possibly a cover-up..." ibid.

"Some of the ice ships had specially built elevators..." Nicholas Pavloff, 1903, as cited in G. Stevens 1986B.

"Initially, the RAC planned to cut ice in Sitka..." Keithahn (1945: 125). The history of the American Russian Company was reported by Keithahn in 1945.

"Ice cutting began in December..." G. Stevens (1986B).

"During the ice-cutting season..." The labor dispute was first described by Captain D. B. Walker, and again by G. Stevens (1986C).

"The round trip from Sitka to Kodiak..." Anichenko and Rogers (2004: 23).

"By 1859, the price for ice had declined..." Keithahn (1945: 123).

"In the early 1860s Russia entered into discussions..." The importance of the Woody Island ice operation to the purchase price for Alaska is told by several sources, including Keithahn, (1945: 123), Jensen (1975: 77), and Stevens (1986A: 8).

"...to a final price of $7.2 million." As an alternative explanation for this figure, Dave McMahan suggested that the additional $200k was intended for bribes to US Congressmen. Correspondence between some of the Russians involved suggests that there might have been something to this. The Senate had approved the Alaska purchase, but Congress was balking at appropriating funds to complete the purchase and needed a little persuasion. Dave believed that only $7 million was actually transferred to Russia.

"The *Aleutian* is of minor importance..." *Anchorage Daily News*, August 24, 2003. "Divers seek rights to a shipwreck 'frozen in time' off Kodiak Island." AP Wire story.

Chapter 14

"Apparently the lawyer Peter Hess..." Hess's talk was reported by the *Anchorage Daily News*, December 9, 2003. See Appendix B.3.

"I spent the next day writing a long letter to the editor." The original text of my letter is included as Appendix B.4. The published version was somewhat shorter.

"I didn't have long to wait for a response." *Kodiak Daily Mirror*, December 16, 2003.

Chapter 15

"On the front page of the *Anchorage Daily News* today..." "Undersea tangle: Dive company that located shipwreck fights the state over preservation." *Anchorage Daily News*, June 10, 2004.

Chapter 18

"...a carefully designed set of sequential hypotheses..." These are described in detail in Cantelas et al. (2005:35).

Chapter 19

"I returned home that evening..." *Kodiak Daily Mirror*, Friday, July 16, 2004, "Grail confirms *Kad'yak* wreck site" by Drew Herman.

"It must have been a big seller." In fact, *Ships' Bilge Pumps* (Oertling, 1996) is available from Amazon Books, and is listed as #379 in Marine Engineering.

Chapter 20

Most of the material in this chapter, including descriptions of the site and artifacts, is redistilled from Cantelas et al. (2005), and Anichtchenko and Rogers (2007).

Chapter 21

The story of the *Big Valley*'s sinking was pieced together from news articles in the *Anchorage Daily News* and *Kodiak Daily Mirror*, and my personal conversation with survivor Cache Seel.

"No sign of crabbers missing at sea" *Anchorage Daily News*, January 17, 2005. Written by Lisa Demer.

"Spokesman of survivor gives details of *Big Valley* tragedy." *Kodiak Daily Mirror*, Tuesday, January 18, 2005. Written by Jan Danelski.

"Weeks later an official inquiry was held..." One report in the *Los Angeles Times*, February 2, 2015, "Load Cited in Sinking Inquiry" stated that the boat was carrying 55 pots, which is apparently in error.

LIST OF ABBREVIATIONS

ADN	*Anchorage Daily News* (now *Alaska Dispatch News*)
ASA	Abandoned Shipwreck Act (of 1987)
KDM	*Kodiak Daily Mirror*
KFRC	Kodiak Fisheries Research Center
KMXT	Kodiak Public Broadcasting Station (Radio)
NRHP	National Register of Historic Places
NMFS	National Marine Fisheries Service
NOAA	National Oceanic and Atmospheric Administration
NURP	National Undersea Research Program
RAC	Russian-American Company
WHOI	Woods Hole Oceanographic Institute

SOURCES

THIS BOOK IS MEANT TO be an entertaining story, and not an exhaustive historical document. For this reason, I do not refer to many of the original sources of information but rather to summaries and secondary sources which are easier to find and assimilate. Much of the history of the *Kad'yak* was included in Evgenia Anichenko's MS thesis at ECU, which may be unavailable to the average reader. Look there for many of the primary sources.

100th Congress of the United States, 1987. *Abandoned Shipwreck Act of 1987.* Public law 100-298 (43 USC 2101).

Anichenko, E. 2005. *Ships of the Russian-American Company, 1799-1867.* MA Thesis. East Carolina University, Greenville, NC.

Anichenko, E., and J. Rogers. 2007. *Alaska's Submerged History: The Wreck of the* Kad'yak. State of Alaska, Department of Natural Resources, Office of History and Archaeology, 550 W. 7th St., Suite 1310, Anchorage, AK 99501.

Arkhimandritof, I. 1860. *Journal of Survey of Russian Skipper Arkhimandritof.* Microfiche copy, Kodiak Library.

Black, L. T. 2004. *Russians in Alaska, 1732-1867.* University of Alaska Press, Fairbanks, AK. 328 pp.

Bowditch, N. 1802. *The American Practical Navigator.* National Imagery and Mapping Agency, Pub. No. 9. 879 pp.

Cantelas, F., T. J. Runyan, E. Anichtchenko, and J. Rogers. 2005. Exploring the Russian-American Company Shipwreck *Kad'yak*. Report for the NOAA Office of Ocean Exploration, Grant # NA04OAR4600043, and National Science Foundation Grant # OPP-0434280. Maritime Studies Program, East Carolina University.

Curryer, B. N. 1999. *Anchors: An Illustrated History*. Naval Institute Press, Annapolis, MD. 160 pp. [Listed on some sites as "Anchors: The Illustrated History"]

Desmond, C. 1919. *Wooden Ship-Building*. Rudder Publishing Company, New York, NY. 240 pp.

Golovin, P. N. 1979. *Civil and Savage Encounters; The worldly travel letters of an Imperial Russian Navy Officer, 1860-1861*. Translated and annotated by Basil Dmytryshyn and E. A. P. Crownhart-Vaughan. The Press of the Oregon Historical Society, Portland, OR.

Jensen, R. J. 1975. *The Alaska Purchase and Russian-American Relations*. University of Washington Press, Seattle, WA. 185 pp.

Keithahn. E. L. 1945. "Alaska Ice, Inc." *Pacific Northwest Quarterly*. 36(2): 121-131.

Lever, D. 1998. *The young sea officer's sheet anchor, or a key to the leading of rigging and to practical seamanship*. Originally published 1819. Dover Publications, Inc., Mineola, NY.

Oertling, T. J. 1996. *Ships' Bilge Pumps: A History of Their Development, 1500-1900*. Texas A&M University Press, College Station, TX. 130 pp.

Pierce, R. A. 1983. *Record of maritime disaster in Russian America, Part two: 1800-1867*. Pp 59-71 in S. Langdon (Ed.), Proceedings of the Alaskan Marine Archaeology Workshop, May 17-19, 1983, Sitka, Alaska. Alaska Sea Grant Report 83-9.

Pierce, R. A. 1990. *Russian America: A bibliographic dictionary*. Limestone Press, Ontario, Canada. 575 pp.

Quinth, S. 2013. *Kodiak, Alaska – the Island of the Great Bear*. Published by Camera Q, Kodiak, AK. Translated from Swedish by LaVonne Quinth. 270 pp.

Rogers, J.S., E. Anichtchenko, and J.D. McMahan. 2008. *Alaska's Submerged History: The Wreck of the* Kad'yak. Alaska Park Science. 7(2):28-33. U.S. National Park Service, Alaska Regional Office, Anchorage.

Scammon, C. M. 1874. *The Marine Mammals of the Northwestern Coast of North America*. Dover Publications, New York, NY. 319 pp.

Slater, R. A. 2012. *Views from the Conning Tower: the Adventures of a Deep-Sea Explorer*. Self-published.

Stevens, B. G. (Editor). 2014. *King Crabs of the World: Biology and Fisheries Management*. CRC Press (Taylor and Francis), Boca Raton, FL. 608 pp.

Stevens, B. G. 1990b. "Temperature-dependent growth of juvenile red king crab *Paralithodes camtschatica*, and its effects on size-at-age and subsequent recruitment in the eastern Bering Sea." *Can. J. Fish. Aquat. Sci.* 47(7):1307-1317.

Stevens, B. G., J. A. Haaga, and W. E. Donaldson. 1994. "Aggregative mating of Tanner crabs *Chionoecetes bairdi*." *Canadian Journal of Fisheries and Aquatic Sciences*. 51:1273-1280.

Stevens, B. G., W.E. Donaldson, J. A. Haaga, and J. E. Munk. 1993. "Morphometry and maturity of paired Tanner crabs, *Chionoecetes bairdi*, from shallow- and deepwater environments." *Canadian Journal of Fisheries and Aquatic Sciences*. 50:1504-1516.

SOURCES

Stevens, B. G., J. Haaga, and W. E. Donaldson. 2000. "Mound formation by Tanner crabs (*Chionoecetes bairdi*): Tidal phasing of larval launch pads?" *Crustacean Issues* 12:445-456.

Stevens, G. 1986A. "Woody Island Ice Company. Part I." *Kadiak Times*, October 9, 1986.

_____. 1986B. "Woody Island Ice Company. Part II." *Kadiak Times*, October 17, 1986.

_____. 1986C. "Woody Island Ice Company. Part III." *Kadiak Times*, October 24, 1986.

_____. 1986D. Woody Island Ice Company. Part IV. Kadiak Times, October 31, 1986.

_____. 1990. *The Woody Island Ice Company*. Russia in North America: Proceedings of the 2nd International Conference in Russian America, 1987. R. A. Pierce (Ed). Sitka, Alaska. Limeston Press, Kingston, Ontario. pp 192-212.

Tebenkov, M. D. 1852. *Atlas of the Northwest Coasts of America, from Bering Strait to Cape Corrientes, and the Aleutian Islands, with several sheets on the Northeast Coast of Asia*. Translated and edited by R. A. Pierce, Limestone Press, Kingston, Ontario, 1981.

Tikhmenev, P. A. 1978. *A History of the Russian American Company*. Originally published in 1864. Translated and edited by R. A. Pierce and A. S. Donnelly. University of Washington Press, Seattle, WA.

Underhill, H. A. 1988. *Sailing Ship Rigs & Rigging, 2nd ed*. Brown, Son & Ferguson, LTD., Glasgow. 126 pp.

Walker, Captain David. B. *Journal and Papers, 1864-1869*. American Russian Commercial Company, Superintendent at Wood Island, Alaska (Ice Works). Alaska State Library Historical Monograph, No. 8. Juneau, AK.

Weightman, G. 2003. *The Frozen Water Trade: A True Story*. Hyperion Press, New York, NY.

WEBSITES

http://alaskashipwreck.com/shipwrecks-a-z/alaska-shipwrecks-a/
The website for alaskashipwreck.com. A compendium of information on shipwrecks in Alaska. I do not know the origin of the information, and therefore cannot verify its provenance. Last accessed, January, 2017.

http://archive.archaeology.org/online/features/kadyak/
Website for Archaeology Archive, a Publication of the Archaeological Institute of America. The article "Tracking Down *Kad'yak*" is a summation of interviews with several people including the author. Last accessed, January, 2017.

http://archive.archaeology.org/online/features/kadyak/archaeologists.html
Article "On the Alaskan Seafloor: Diving on the Kad'yak Wreck" features a photo of the *Kad'yak* wreck by Tane Casserly. Last accessed, January, 2017.

http://archive.archaeology.org/online/interviews/stevens.html
Article "From Crabs to Shipwrecks." Interview with the author, August 8, 2004. Last accessed, January, 2017.

https://bgstevens9.wixsite.com/crabman
Academic website of Dr. Bradley Stevens. Repository of information on marine biological research and anything related to crabs. Last accessed, June 2017.

https://deadliestreports.wordpress.com/2007/03/27/two-tragedies-from-season-1-jan-05/
Described as "A Fansite for Discovery's emmy-award winning show 'Deadliest Catch'!" I cannot verify the source of any information listed there, although it includes quotes from an interview with the author (original source not cited).

http://dnr.alaska.gov/parks/oha/projects/kadyak#discovery
Website about the Kad'yak discovery and survey created by Alaska Department of Natural Resources, with input from Dave McMahan. Last accessed 8 February, 2018.

http://oceanexplorer.noaa.gov/explorations/04kadyak/welcome.html
This webpage is the official NOAA account of the survey expedition conducted in 2004, with assistance from the NOAA Ocean Exploration Program. Last accessed, January, 2017.

http://www.cameraq.com/
Website for CameraQ, media publishing company of Stefan and LaVonne Quinth. "Welcome to the Wild and Wondrous World of Camera Q. Documenting Nature and People has been our job since the middle of the '70s." Last accessed, June 2017.

https://www.greeka.com/dodecanese/kalymnos/sponge-diving-tradition.htm and http://www.divingheritage.com/greecekern2.htm
Both of these sites relate the history of Greek sponge diving in the Mediterranean. Last accessed January, 2018.

http://www.noaanews.noaa.gov/stories2004/s2270.htm
A page from NOAA magazine, telling a similar story to the OEP webpage. Last accessed, January, 2018.

https://www.padi.com/
Website for PADI, the Professional Association of Diving Instructors, with information on scuba dive training and requirements. Last accessed, January, 2018.

https://www.omao.noaa.gov/learn/diving-program
Website for the NOAA Diving Program. Last accessed, January, 2018.

https://www.aaus.org/
Website for the American Academy of Underwater Sciences. Last accessed, January, 2018.

ABOUT THE AUTHOR

BRADLEY G. STEVENS IS A marine biologist with a PhD in Fisheries Science from the University of Washington. For twenty-two years, he worked for the National Oceanic and Atmospheric Administration (NOAA) in Kodiak, Alaska, conducting research on king crabs and other species of commercially valuable crabs. Now a tenured Professor of Marine Science at the University of Maryland Eastern Shore (UMES), he serves as Distinguished Research Scientist for the NOAA-funded Living Marine Resources Cooperative Science Center.

In 1999 and 2002, he led cruises to explore Gulf of Alaska Seamounts, diving to over 11,000 feet inside *Alvin*, the same submersible used by Bob Ballard to explore *Titanic*. In 2003–04 he discovered and helped survey the wreck of the Russian barque *Kad'yak* (1860), the only shipwreck from the Russian Colonial Period and the oldest known wreck site in Alaska. Stevens has authored over sixty scientific research articles, and in 2014 published *King Crabs of the World: Biology and Fisheries Management*, the only comprehensive reference book about king crabs. He is an accomplished rock and jazz drummer, and has been sighted in various Irish bars playing Celtic music on mandolin and hammered dulcimer. When not working, he spends his time sailing or kayaking. He lives with his wife Meri Holden.

CPSIA information can be obtained
at www.ICGtesting.com
Printed in the USA
BVHW05s0736040818
523124BV00002B/2/P

9 781513 261379